Narendra Pal Singh Chauhan, Narendra Singh Chundawat
**Inorganic and Organometallic Polymers**

# Also of interest

*Two-Component Polyurethane Systems*
Chris Defonseka, 2019
ISBN 978-3-11-063957-5, e-ISBN (PDF) 978-3-11-064316-9,
e-ISBN (EPUB) 978-3-11-063979-7

*Recycling of Polyethylene Terephthalate*
Martin J. Forrest, 2019
ISBN 978-3-11-063950-6, e-ISBN (PDF) 978-3-11-064030-4,
e-ISBN (EPUB) 978-3-11-064045-8

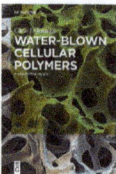

*Water-Blown Cellular Polymers*
Chris Defonseka, 2019
ISBN 978-3-11-059784-4, e-ISBN (PDF) 978-3-11-064312-1,
e-ISBN (EPUB) 978-3-11-063978-0

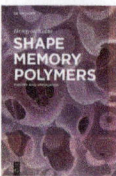

*Shape Memory Polymers*
Hemjyoti Kalita, 2018
ISBN 978-3-11-056932-2, e-ISBN (PDF) 978-3-11-057017-5,
e-ISBN (EPUB) 978-3-11-056941-4

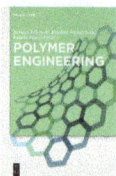

*Polymer Engineering*
Bartosz Tylkowski, Karolina Wieszczycka, Renata Jastrząb (Eds.), 2017
ISBN 978-3-11-046828-1, e-ISBN (PDF) 978-3-11-046974-5,
e-ISBN (EPUB) 978-3-11-046834-2

Narendra Pal Singh Chauhan,
Narendra Singh Chundawat

# Inorganic and Organometallic Polymers

—

DE GRUYTER

**Authors**
Narendra Pal Singh Chauhan
Department of Chemistry
Bhupal Nobles' University
Maharana Pratap Station Road
313001 Udaipur, Rajasthan, India
narendrapalsingh14@gmail.com

Narendra Singh Chundawat
Department of Chemistry
Bhupal Nobles' University
Maharana Pratap Station Road
313001 Udaipur, Rajasthan, India
chundawat7@yahoo.co.in

ISBN 978-1-5015-1866-9
e-ISBN (PDF) 978-1-5015-1460-9
e-ISBN (EPUB) 978-1-5015-1479-1

**Library of Congress Control Number: 2019950281**

**Bibliographic information published by the Deutsche Nationalbibliothek**
The Deutsche Nationalbibliothek lists this publication in the Deutsche Nationalbibliografie;
detailed bibliographic data are available on the Internet at http://dnb.dnb.de.

© 2019 Walter de Gruyter GmbH, Berlin/Boston
Typesetting: Integra Software Services Pvt. Ltd.
Printing and binding: CPI books GmbH, Leck
Cover image: LAGUNA DESIGN / Science Photo Library /
Getty Images

www.degruyter.com

# Contents

# Introduction

Inorganic polymers with an inorganic backbone and organometallic groups as pendant groups are one of the most promising approaches to new materials that combine the advantages of organic polymers with those of inorganic solids. According to the IUPAC Compendium of Chemical Terminology V2.3.3, inorganic polymers are polymers or polymer networks with a skeletal structure that does not include carbon atoms. Organometallic polymers are the entities where the repeating units of a metal attach to a chemical bond with carbon atoms of an organic molecule. Various desired properties can be introduced using different backbone elements and side groups, and methods of syntheses have been briefly outlined.

The aim of this book is to present to the students an introduction to the developments in inorganic and organometallic polymers with general overview on synthesis, characterization, and applications of inorganic and organometallic polymer synthesis. This book is divided into eight chapters. Chapter 1 includes silicon-based polymers such as polycarbosilane and polysiloxane, and Chapter 2 includes tin- and germanium-based polymers such as polystananes and polygermanes with their properties. Chapter 3 reveals about various preparation methods of polyphosphate, polyphosphoric acids, phosphonate and polyphosphazene, whereas sulfur-based polymers like polythiazyl, parathiocyanogen, polythiol, polysulfur, polysulfide, and polysulfone are included in Chapter 4. Chapter 5 deals with synthetic methods and applications of organometallic polymers. Various coordination and geopolymers are discussed in Chapters 6 and 7, respectively. Chapter 8 deals with various advanced characterization, including spectral and thermal microscopy such as scanning electron microscopy, atomic fluorescence microscopy, transmission electron microscopy, and also single-crystal X-ray diffraction analysis.

We believe that this book would be of general interest to organic chemists, pharmacists, physicists, polymer scientists, food scientists, and technologists. We earnestly hope this book offers a balanced, interesting, and innovative perspective in the area of both academic and industrial point of view.

## Conflict of interest

The author(s) declared no conflicts regarding each of the chapters of this book.

https://doi.org/10.1515/9781501514609-001

**Acknowledgments:** The authors would like to thank the publisher De Gruyter. We would also like to thank Dr. Helene Chavaroche, Editorial Manager Publications for her valuable contributions. We extend our heartfelt and deep gratitude to our workplace Bhupal Nobles' University, Udaipur.

Dr. Narendra Pal Singh Chauhan
Dr. Narendra Singh Chundawat

# 1 Silicon-based polymers: polysilane, polycarbosilane, polysiloxane, and polysilazanes

**Abstract:** In this chapter, silicon-based inorganic polymers have been discussed. The silicon polymers have a great variety of commercial applications such as resins, elastomers and fluids. Silicon is a quickly oxidized and cross-linkage-type polymer. In general, silicon-based polymers have low molecular weight and are elastic in nature. Also, synthetic method, properties, and its important role in various commercial industries are discussed in detail.

**Keywords:** silicon-based polymers, siloxane, synthesis, glass transition temperature

## 1.1 Introduction

Silicon-based polymers generally show the properties such as superior chemical resistance along with high thermal stability and also resistance to UV radiations at low temperatures. Due to these vital properties, siloxane coatings are widely used as industrial organic coatings, applied to structures and instruments [1]. Pre-ceramic polymers are generally made up of silicon through the pyrolysis technique [2]. As an outcome, they play a vital role in the application part such as ceramic barrier coatings [3, 4], high-thermal stabilized materials, and controlled porous membranes from top to bottom [5–8].

Out of various pre-ceramic polymers, silazanes show good results in coating applications, due to the alternate arrangement of silicon and nitrogen on the polymers they sense moisture (water molecule) from the atmosphere and if silazane carries Si–H groups, then it helps in improving the adhesion between silazane and surfaces like metals, glass, plastics, and ceramics. Because of more presence of –OH groups [9,10], they develop strong substrate–O–Si bonds [11].

Polycarbosilanes (PCSs) are also known as organosilicon polymers and its backbone is based on silicon atoms, in proper proportion bifunctional organic groups coordinate to silicon atoms. Most of the leading silicon carbide (SiC) materials have various structural and functional applications due to their thermomechanical properties [12].

PCS is prepared by thermal decomposition of polydimethylsilane in an autoclave under high pressure [13]. To avoid an expensive autoclave method, Yajima et al. [14] prepared PCS without using an autoclave; thus, a low-molecular-weight ceramic product was obtained. Polysilazanes (PCSNs) are the simplest silicon units containing a polymer; silicon nitride ($Si_3N_4$) and SiC are used as precursors for the facile way to synthesize.

https://doi.org/10.1515/9781501514609-002

High elastic modulus along with the properties of great tensile strength and high thermal stability material is SiC fibers, which is employed in cement as well as forced ceramic–matrix composites [15, 16]. SiC is applicable as ceramic coating due to the superb properties of toughness and hardness, which is used to fill the scratches on the glass and metal type of materials [17–19]. Nowadays, it is very popular in a field of high-temperature electronic devices along with optoelectronic devices for future generation point of view [20].

Polycrystalline β-SiC films especially known as amorphous in nature are considered for the applications such as semiconductor, hard protective coatings and structural materials for power microelectromechanical systems in harsh environments. SiC films are auspicious in various applications due to their electron saturation drift velocity, large energy gap, high mobility, high electric breakdown field, high thermal conductivity, chemical inertness and thermal stability [21]. SiC films show luminescent materials that emit in the visible spectrum ranging from blue to yellow [22]. SiC coatings on substrates' facile way were employed by using physical vapor deposition, chemical vapor deposition (CVD), and sputtering and molecular beam epitaxy [19–22]. SiC/substrate creates a problem in the coating during the starting stage of the development due to structural defects. Hence, this is a very hot topic in the SiC-based microelectromechanical system devices [23].

Nonwoven PCS fibers produced by electrospinning are used for many purposes such as enhanced filler in a composite material [24], acting as a filtration membrane [25], nanosensor and enzyme immobilization [26]. Processing the PCS fiber with curing oxidation, electron beam irradiation [27], or γ-ray [28] will produce infusible fibers to produce a high yield and to improve the properties of the ceramics processing with pyrolysis method.

The alternate arrangement of silicon–oxygen combination chain on the inorganic polymer is called polysiloxanes [29, 30]. The chemical formula of polysiloxanes is $(R_2SiO)_n$, where R is usually methyl siloxanes ($CH_3$), although it can also be H or alkyl or aryl group. The poly(dimethylsiloxane) (PDMS) is the most ample studied polymer, which consists of methyl groups as the substituents on silicon.

## 1.2 Synthetic methods

### 1.2.1 Polydimethylsilane

Polydimethylsilane is prepared by using dimethyldichlorosilane [$SiCl_2 (CH_3)_2$] with sodium in xylene at ~120 °C temperature [31]. Metallic sodium was chopped into 5 mm × 5 mm pieces and introduced into the reaction flask containing xylene under a nitrogen atmosphere; metallic sodium completely melts at 98 °C and moves to the bottom of the flask. Chlorination takes place after the addition of dimethyldichlorosilane drop by drop into the same flask with constant stirring. Refluxing the reaction mixture for 8 h at 120 °C, xylene has been removed from the reaction mixture by using suction filtration technique; after natural drying bluish purple precipitate has been obtained. Methanol has been added to remove the unreacted sodium; in continuation polydimethylsilane and NaCl were removed by deionized water and acetone.

## 1.2.2 Polycarbosilane

The polydimethylsilane powder is added in a round-bottomed flask along with a condenser under nitrogen gas atmosphere. The polydimethylsilane [$SiCl_2(CH_3)_2$] powder heated in the furnace, fluxed the reaction mixture at 370 °C, turned to creamy white.

The polydimethylsilane melted and then turned into a liquid state at 370 °C, and by refluxing the reaction mixture up to ~450 °C, volatile components expelled out easily. In the viscous product, n-hexane has been added and the reaction mixture has been stirred, followed by filtration. The yellowish-brown-colored viscous liquid obtained, n-hexane, can be separated from the final product by using a rotary evaporator and then lustrous yellowish crystals obtained as a final product.

The PCS was prepared by using Poly(dimethylsiloxane) (PDMS) at normal pressure [32]. In general, PCS with composition (Si: 41.03 wt%, O: 1.05 wt%, C: 43.24 wt%) were synthesized, and transparent solid material would be obtained with number average molecular weight ($M_n$) of 1426 DA and a weight average molecular weight ($M_w$) of 3296 DA. The PCS polymer shows a melting point at 215 °C [33, 34].

## 1.2.3 Polysilazanes

Seyferth and coworkers [35] developed polyorganosilazane by using ammonolysis reaction method, as a precursor dichlorosilane ($H_2SiCl_2$) and ammonia ($NH_3$) in polar solvents (ether and dichloromethane) shows remarkable properties like extremely high thermal stability. Xie and Li [36] synthesized long-chain PCSN by using dichloromethylsilane (MeHSiCl$_2$) instead of $H_2SiCl_2$ and ammonia along with catalyst ammonium chloride (NH$_4$Cl). Wang and coworkers [37] have reported a preparation of ethynyl-terminated PCSN from the ammonolysis reaction of dichloromethylsilane (MeSiHCl$_2$) and p-phenylenediamine, by which terminal ethynyl groups were introduced into the resultant PCSN (Figure 1.1). Seo and coworkers [38] have fabricated composite using PCSNs, which is a highly transparent composite material and its refractive index depends on diphenylsilanediol.

Amorphous nature along with nonoxide ceramic material prepared by employing SiBNC (derived from polyborosilazane), SiC (derived from polyborazine), and BN (derived from PCS) [39–43], the SiBNC network structure shows low density, excellent oxidation resistance, and thermal as well as chemical stability; and thermodynamic machines have also drawn more attention to the same material [44, 45].

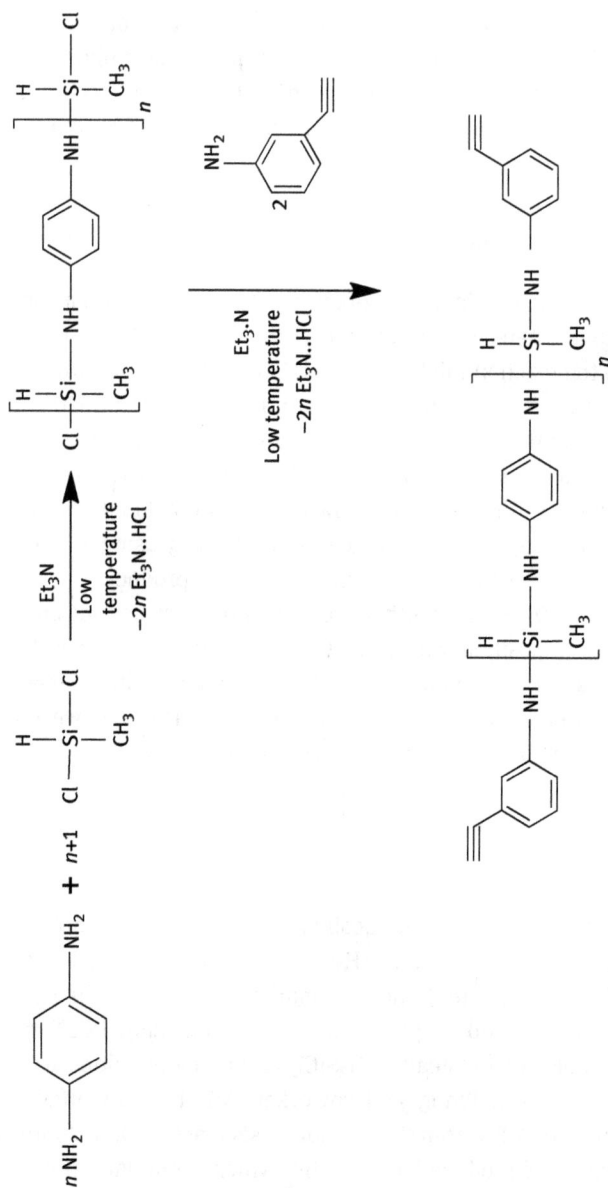

**Figure 1.1:** Synthesis scheme of PCSN. PCSN – polysilazane.

Furtat et al. [46] have synthesized fluorine-based PCSNs by using commercially available oligosilazane Durazane 1800, and different amounts of 2,2,2-trifluoroethanol with Tetra-n-butyl-ammonium fluoride (TBAF) as a catalyst were developed for their application as low surface energy coatings with excellent adhesion to protect metal substrates.

The thermosetting hybrid-based PCSN resin modified by applying organic PCSNs with various organic polymers, cross-linked with starting materials in the form of a coating [47] as well as redesigned PCSN thermosetting hybrid resins based on polycyanurates [48] were reported. PCSNs have more reactivity toward oxygen, moisture, peroxides, and acids due to almost no difference in standard enthalpy between oxygen and silicon. The direct combinations are between PCSNs and unsaturated polyesters as matrix resins for FRPs [49].

### 1.2.4 Polysiloxanes

The condensation of two silanols:

$$2\,R_3Si - OH \rightarrow R_3Si - O - SiR_3 + H_2O$$

Generally, the silanols are synthesized in situ by silyl chlorides hydrolysis. With a disilanol, $R_2Si\,(OH)_2$, the condensation can generate linear products with silanol groups at terminals:

$$n\,R_2Si(OH)_2 \rightarrow H(R_2SiO)_n OH + nH_2O$$

Another way, the disilanol can afford cyclic products

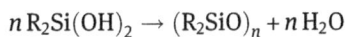

$$n\,R_2Si(OH)_2 \rightarrow (R_2SiO)_n + n\,H_2O$$

Polysiloxanes are prepared by the following two methods:
(a)  By ring-opening polymerization of cyclosiloxanes
(b)  By condensation polymerization with a hydrolysis/condensation reaction of di-organodichlorosilanol condensation reaction between two difunctional diorganosilanes (homocondensation). Both of these reactions are efficient routes of forming polysiloxanes; however, commonly studied method is homocondensation. Homocondensation, which is also known as polycondensation, is a reaction in which two molecules, including the silanol group, condense to form a polymer. This is shown in the following reactions:

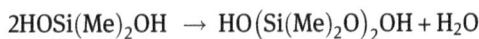

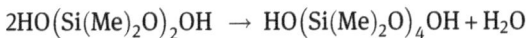

$$2HOSi(Me)_2OH \rightarrow HO(Si(Me)_2O)_2OH + H_2O$$

$$2HO(Si(Me)_2O)_2OH \rightarrow HO(Si(Me)_2O)_4OH + H_2O$$

The siloxane can activate the hydroxyl group of monomer that is responsible for high reactivity of the molecule in polycondensation [50]. These reactions are as follows:

〜〜 SiOSiOH + HOSi 〜〜 ⟶ 〜〜 SiOSiOSi + $H_2O$

〜〜 SiOSiOH + HOSi 〜〜 ⟶ 〜〜 SiOSiOH + HOSiOSi 〜〜

The result reveals that the geometry of the orbitals and their interactions stabilizes the silanol structure and strengthens the silicon to hydroxyl group bond. The molecular interactions between molecules are high because the silanol group is both a good proton donor and acceptor. Because of this, the silanol-to-silanol intermolecular and intramolecular hydrogen bonds are readily formed. For polycondensation, these interactions can be considered as catalysts. The silanol end group is a very effective catalyst for short chains; this becomes a less effective catalyst with elongation of the chain due to the siloxane chain length.

An alternate way to form polysiloxanes is the ring-opening reaction that can form either block or random copolymers. The polymerization occurs only if a few of the comonomer is in solution. Long chain length and the high degree of polymerization occur with the low amount of comonomer. A generic equation for this reaction is

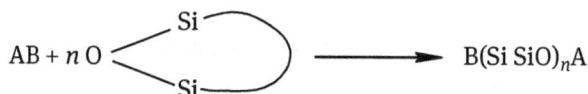

$$AB + n\, O\!\!\left\langle\!\!\begin{array}{c} \text{Si} \\ \text{Si} \end{array}\!\!\right) \longrightarrow B(Si\ SiO)_n A$$

where the initiator, AB, can be either an acid or a base. Few initiators are listed below:

The cyclosiloxane polymerization reaction can be done using the following initiators:

Basic initiators: XOH, XOSi = , XOR, XR, XSR

Acidic initiators: $HI_3$, $HCl-FeCl_3$, $CF_3SO_3H$, HF, $H_2SO_4$, where X is an alkali metal or phosphonium group or quaternary ammonium and R is an alkyl, polystyryl, or poly(trimethylsilylvinyl). Use of a strong acid or base results in reversible reaction and the equilibrium constant is represented as

$$k = [SiOSi][H_2O]/[SiOH]^2$$

New highly functioned polysiloxanes behaves like organometallic. It consists of two distinct redox active (oxidizing and reducing) types: the electron-withdrawing $(\eta^6\text{-aryl})$tricarbonylchromium and the electron-donating ferrocenyl group [51].

The organometallic compound like ferrocenylpolysiloxanes is a most soluble organic solvent; it is more stable in air and humidity. Thermal stability of the novel ferrocenyl polysiloxanes strongly depends on the size of the ferrocenyl dendritic fragment appended to the siloxane backbones. Ferrocenyl dendronized polysiloxanes are thermally stable at 200–250 °C under $N_2$ and at more elevated temperatures yields ceramic residues in relatively high amounts.

Regular silsesquioxanes, those with ladder structures, have got considerable attention. Data from XRD, $^{29}$Si NMR spectra, and IR under certain conditions have pointed out the formation of ladder-type polysilsesquioxanes. Si–Si bonds are oxidized by mCPBA; as a result, ladder-type morphology is observed (Figure 1.2) [52].

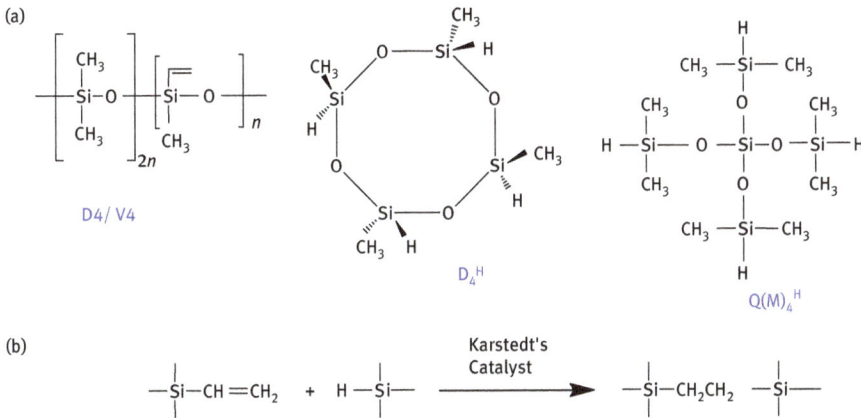

(a)

D4/ V4

$D_4{}^H$

$Q(M)_4{}^H$

(b)

**Figure 1.2:** Synthesis of polysiloxane networks: (a) polysiloxane $D_4/V_4$ and cross-linking agents – $D^H{}_4$, $Q(M^H)_4$; (b) hydrosilylation of vinyl compounds.

Hydrolysis of RR'SiCl$_2$ generates diorganosilanediol, and the self-condensation technique helps to develop different categories of products such as linear and cyclic. The following reactions show a continuous process:

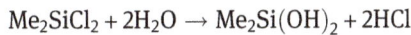

$$Me_2SiCl_2 + 2H_2O \rightarrow Me_2Si(OH)_2 + 2HCl$$

$$Me_2Si(OH)_2 > [Me_2SiO]_n + nH_2O$$

The above reaction shows linear polymeric in nature with lesser dilution due to quite sensitive natural conditions. A catalyst basically ends to linear products with high molecular weight. Alternatively, it is found that employing an acidic catalyst tends to incline toward the formation of cyclic light-weight polymer products. There are various ways to synthesize bifunctional silicon-type derivatives, which could be used for step-growth polymerization. For example, diaminosilanes are suitable synthons for diorganosilane-diols condensation, which is easiest due to the vibrant Si–N bond. Thus, the reaction of Me$_2$Si(NMe$_2$)$_2$ with diphenylsilanediol results in a random copolymer [53], due to the catalytic action of the released amine to split Si–O bonds to randomly obtain

$$Me_2SiCl_2 + LiNMe_2 ----------> Me_2Si(NMe_2)_2$$

By allowing bisureidosilane to condense with diphenylsilanediol alternating copolymers can be synthesized. With stepwise way, initially pyrrolidine along with $Me_2SiCl$ the monomer bis-ureidosilane is synthesized. The pyrrolidino derivative on reaction with PhNCO gives the bis-ureidosilane. Hence, the condensation reaction of the diphenylsilanediol with bis-ureidosilane affords the by-product urea in the form of precipitate that emerged from the reaction mixture, interchanging he copolymer containing $[Me_2Si-OSiPh_2O]_n$.

Most vital cross-linking polysiloxanes are employed as advanced elastomers. Theoretically, it is quite facile but its technology is quietly disrupted. Preserving the outlook of this book in mind, bypass the technological discussion. Reinforcing additive such as fiber-type high-surface-area silica high-molecular-weight polysiloxanes are heated with each other along with organic peroxides and some coloring pigments to afford cross-linked products via $-CH_2CH-$ links.

To the nature of the reaction conditions, this process is quite sensitive. Linear polymeric products are formed by using less dilution. High-molecular-weight linear products are favored with the use of basic catalysts. But when acidic catalysts are used, they favor the formation of lightweight cyclic product. Bifunctional silicon derivatives like polysiloxanes can be utilized for step-growth polymerization by applying hydrolysis. For example, diaminosilanes are known to be good synthons for condensation with diorganosilanediols. Because of the lability of the Si–N bond, the method is very effective. Hence, diphenylsilanediol affords a random copolymer with the reaction of $Me_2Si(NMe)_2$ [54].

Simultaneously, copolymers can be synthesized by choosing bis-ureidosilane, followed by condensation with diphenylsilanediol. Thus, the condensation reaction of the bis-ureidosilane with diphenylsilanediol precipitates out of the reaction mixture and therefore does not participate in further reactions. Alternatively, copolymer containing $[Me_2Si-OSiPh_2O]_n$ is produced. Chojnowski et al. [55] prepared a branched polysiloxane by anionic ring-opening polymerization (Figure 2.2).

The linear polymer chain can be obtained by the most effective ring-opening polymerization of cyclosiloxanes. The anionic polymerization through ring opening of heptamethyl-1,3-dioxa-5-aza-2,4,6-trisilacyclohexane with organolithium compounds, dimethylformamide (DMF) solvent, is used to control polymerization [56]. Cyclosiloxanes show an alternate arrangement of silicon and oxygen atoms in cyclic structures known as inorganic heterocyclic rings [57], and tetravalent silicon carries two substitutes and oxygen bi-coordinate. Cyclosiloxanes are near about the same toward cyclophosphazenes. Unlike cyclophosphazenes, cyclosiloxanes do not contain a valence unsaturated skeleton.

Various types of cyclosiloxanes comprising especially aryl and alkyl substituents on the silicon are well known, and six- and eight-membered ring compounds are mostly observed. In hydrolysis reaction of the trichlorosilane, three-dimensional silsesquioxanes are obtained. The ring sizes of cyclosiloxanes are quite variable. The four-membered ring is about the smallest size, while larger rings can also form as cyclic oligomers with $a$ = 4–23 [58]. The diorganosilene diol can be prepared by RR'SiCl$_2$ hydrolysis, which undergoes self-condensation resulting in a different type of cyclic and linear products. Ring-opening polymerization of cyclosiloxane involves the following steps:

1.  Ring of cyclosiloxanes can be opened through the polymerization method by using cationic or anionic initiators.
2.  Either by protic or Lewis acids, polymerization takes place with cationic initiators . Mechanisms of different types of polymerization are recommended such as acidolysis/condensation, polymerization via silylium ion-mediated or oxonium ion-mediated polymerization.
3.  The high-molecular-weight polymer can be generated by metal hydroxides, and anionic polymerization can be achieved even through lithium salts of silanols.
4.  Methyl-Si groups are used to stop the polymerization reaction and form stable products.
5.  Copolymers can be synthesized by polymerization of either a mixture of cyclosiloxanes or heterogeneous substituted cyclosiloxanes.

Polysiloxanes have high biocompatibility and are less harmful due to these properties, thereby creating an important role in the polymer world and expanding utility in various fields. Ning et al. [59] have fabricated on reinforcing PDMS by the in situ precipitation of silica. These were well dispersed at the nanolevel, and considerable reinforcement of the elastomer could be achieved. Moreover, other particles like titania, zirconia, alumina, and silica were also employed to have a good reinforcement on elastomers [60, 61]. Various synthetic approaches and applications are given in Table 1.1.

## 1.3 Properties

Synthesized polymers consist of the functional group silanol, SiO(Me)$_2$, and polysiloxanes can be in forms of oils, greases, rubbers, or plastics. In order to understand the characteristics of polysiloxane polymers, it is important to understand the chemistry of individual elements of the polymer and the behavior of the functional group. Few properties have resemblance with carbon, but on the whole, it is completely a different element. It constitutes around 27% of the Earth's crust and it is second in abundance in the Earth after oxygen. It is very rare to find silicon in elemental form in nature, and is found usually in the bounded form of oxygen as either SiO$_2$ or SiO$_4^{2-}$. Silicon bonding is quite similar to carbon bonding in many ways. Carbon is the main content of the living world and can form infinite length chains. Silicon oxides are more stable than silicon–silicon bonding; chains are more readily made from silicon oxide, resulting in siloxanes. Polysiloxanes are among the most flexible polymers known because of many factors like long skeletal bond lengths, wide angles as oxygen, and poor intermolecular interactions.

**Table 1.1:** Various synthetic approaches and applications.

| S. no. | Synthetic approach | Application | Reference |
|---|---|---|---|
| 1 | Polysiloxane bearing quaternary ammonium and *N*-halamine | Antimicrobial coating | [62] |
| 2 | Polysiloxane with trisilanolphenyl–polyhedral oligosilsesquioxane (trisilanolphenyl–POSS) | Optoelectronics | [63] |
| 3 | Polymer electrolytes based on polydimethylsiloxane bearing 1-*N*-methylimidazolium-pentyl iodide side chains | Photochemical | [64] |
| 4 | Polysiloxane/epoxy hybrids with poly glycidyl methacrylate matrix | High-performance coating | [65] |
| 5 | Polysiloxane-based gel agent prepared with poly(dimethylsiloxane-*co*-alkylmethylsiloxane) | Batteries | [66] |
| 6 | Polysiloxane–polyether block a-hydroxyalkylphenones macromolecular photoinitiators with ionic liquid character | Antibacterial agents | [67] |
| 7 | Polysiloxane with ion-conducting poly (ethylene oxide) groups cross-linked with modified gallic acid prepared by thiol-ene click reaction | Electrode for lithium metal battery | [68] |
| 8 | Poly-*p*-phenylenebenzobisoxazole fiber as hyperbranched polymer | Cell engineering | [69] |
| 9 | Poly(siloxane methyl vinyl) doped with Fe clusters/$Al_2O_3$ | EMW shielding | [70] |
| 10 | Poly-(*N*-isopropylacrylamide)/chitosan microgel combined with methyloctadecyl (3-(trimethoxysilyl)propyl) ammonium chloride | Excellent antibacterial agents | [71] |
| 11 | Polysiloxane-modified phenolic resin synthesized by resorcinol and isocyanatopropyltrimethoxysilane, the hybrid resin | Excellent thermal and antioxidant properties | [72] |
| 12 | *N*-Halamine 3-(3-hydroxypropyl)-5,5-dimethylhydantoin (HPDMH) reacted with poly(methylhydrosiloxane) (PMHS) to synthesize polysiloxane with 5,5-dimethylhydantoin-based *N*-halamine pendants | Biocidal properties | [73] |
| 13 | Polyglycerol-modified polysiloxane surfactants synthesized by grafting to approach | Coating application | [74] |

**Table 1.1** (continued)

| S. no. | Synthetic approach | Application | Reference |
|---|---|---|---|
| 14 | Macrocyclic poly(methylvinylsiloxane)s (cPMVSs) synthesized from 1,3,5,7-tetramethyl-1,3,5,7-tetravinylcyclotetrasiloxane with water catalyzed by anhydrous iron(iii) chloride | Excellent thermal stability | [75] |
| 15 | Hydrosilylation reaction of methacryl polyhedral oligomeric silsesquioxane and different hydrosilyl synthesized polysiloxane-based encapsulants | White-light-emitting diodes | [76] |
| 16 | Single epoxy-terminated polydimethylsiloxane (SEPDMS) synthesized by hydrogen-terminated polydimethylsiloxane and allyl glycidyl ether in chloroplatinic acid. Furthermore, dendritic-linear surfactants 1G PAMAM–Si and 2G PAMAM–Si are prepared by grafting of SEPDMS | 2G PAMAM–Si is superior to 1G PAMAM–Si at the ability of emulsifying oil | [77] |
| 17 | Post-polymerization of thioether groups bearing polysiloxanes and the modification of the vinyl groups of poly(methylvinylsiloxanes) with sulfonyl groups via thiol-ene chemistry | Actuators, capacitors, and flexible electronics. | [78] |
| 18 | Silica and phenyltrimethoxysilane/γ-(2,3-epoxypropoxy) propytrimethoxysilane | Excellent waterproof applications because of superhydrophobic nature | [79] |
| 19 | Epoxy-functional urethane and siloxane | Coating applications | [80] |
| 20 | Dendritic polyethylene brushes with polysiloxane as the main chain | – | [81] |
| 21 | The amino-functionalized $NH_2$-HBPSi grafted to PI chains during in situ polymerization, promoting the uniform dispersion of the $NH_2$-HBPSi nanoparticles and forming strong interfacial interactions between $NH_2$-HBPSi and the PI matrix | High-performance reinforcing fibers | [82] |
| 22 | Hybrid of silica nanoparticles and polysiloxane with photo and thermal cross-linkable resin | High thermal stability | [83] |
| 23 | Interpenetrating polymer networks of incompatible poly(dimethylsiloxane) and zwitterionic polymers | Antifouling | [84] |

**Table 1.1** (continued)

| S. no. | Synthetic approach | Application | Reference |
|---|---|---|---|
| 24 | Polysiloxane-modified tetraphenylethene using polycondensation. | Excellent fluorescence quenching efficiency | [85] |
| 25 | Polysiloxane wax groups and polyether amino groups synthesized by hydrosilylation and ring-opening reaction of polyhydromethylsiloxane with stearyl methacrylate, polyether epoxy compound and N-(β-aminoethyl)-γ-aminopropyl methyl dimethoxysilane | Better elasticity and hydrophobicity | [86] |
| 26 | Hybrid inorganic–organic matrix doped with LiTFSI, which consists of ring-like oligo-siloxane clusters, bearing pendant, partially cross-linked, polyether chains | Polymer electrolytes for application in lithium metal batteries | [87] |
| 27 | Pb nanoparticles modified thiol-functionalized polysiloxane film glassy carbon electrode via copolymerization | Determination of Hg(II) | [88] |
| 28 | Polysiloxane functioned with amino acetic acid groups synthesized using sol–gel technology | Better sorption properties | [89] |
| 29 | Cross-linked silicone elastomers constructed with dynamic-covalent boronic esters using thiol–ene "click" chemistry | Self-healing properties | [90] |
| 30 | Polysiloxane material with thiol-amine-based multiple functional groups polysiloxane-(monoamine-thiol) triacetate synthesized | Functional material | [91] |
| 31 | Polysiloxane derivatives having quaternized imidazolium moieties, with different lengths of alkyl chains prepared via quaternization reaction of poly(3-chloropropylmethylsiloxane) with 1-alkylimidazole derivatives | Functional material | [92] |
| 32 | Trifluoropropylmethylsiloxane-phenylmethylsiloxane copolymerized with polysiloxanes | Functionalized material | [93] |
| 33 | Polysiloxane networks obtained via cross-linking of $D_4/V_4$ polysiloxane served as matrices for incorporation of metallic Pt particles | Cross-linking agents | [94] |
| 34 | Cross-linked cyclic siloxane (Si'O) and silazane (Si'N) polymers synthesized via solvent-free initiated chemical vapor deposition | Functionalized material | [95] |

**Table 1.1** (continued)

| S. no. | Synthetic approach | Application | Reference |
|---|---|---|---|
| 35 | Polysiloxane derivatives having quaternized imidazolium moieties, with hydroxyalkyl groups prepared by quaternization of poly (3-chloropropyl-methylsiloxane) using 1-(ω-hydroxyalkyl)imidazole derivatives and anion-exchange reaction using lithium bis (trifluoromethanesulfonyl)imide | Functionalized material | [96] |
| 36 | Polysiloxane side chain liquid crystal polymers with chiral and achiral substitutions in the side chains synthesized via thiol-ene click chemistry | Functionalized material | [97] |
| 37 | Liquid crystalline polymers with polysiloxane backbones | Functionalized material | [98] |
| 38 | Polysiloxane derivative containing 1,3-bis (9-carbazolyl)benzene moiety as a pendant unit on the polysiloxane backbone | Better thermal stability | [99] |
| 39 | Elastic, wearable cross-linked polymer layer tunable polysiloxane-based material | Better mechanical properties | [100] |
| 40 | Hydrogenated rosin (HR) modified vinyl polysiloxane synthesized from HR, hexamethyldisiloxane (MM), 1,3,5,7-tetravinyl-1,3,5,7 tetramethylcyclotetrasiloxane ($D_4Vi$), octamethylcyclotetrasiloxane ($D_4$) and hydrolyzate of γ-aminopropyl(diethoxy) methylsilane | Functional materials | [101] |
| 41 | Polysiloxanes (QPEPSs) with pendant quaternary ammonium polyether groups synthesized having comb-like structure | Antibacterial activity | [102] |
| 42 | Methacrylate-terminated polysiloxane hybrid oligomers and functional acrylate oligomers synthesized | Thermal stability | [103] |
| 43 | Amine-terminated polysiloxane and succinic anhydride-based carboxyl-terminated polysiloxane | Excellent fluorescence properties. | [104] |
| 44 | Phosphate, borates, bromides, sulfonate, and imide-based polymeric ionic liquid crystals | Functional material | [105] |
| 45 | Amino-grafted polysiloxane surfactants with well-defined amphiphilic structures synthesized | Functional material | [106] |

**Table 1.1** (continued)

| S. no. | Synthetic approach | Application | Reference |
|---|---|---|---|
| 46 | Triphenylamine-based polysiloxane synthesized. | Functional polymers | [107] |
| 47 | Comb-like QPEPS bearing dimethyl dodecyl quaternary ammonium polyether groups synthesized | Antibacterial activities | [108] |
| 48 | 1D polysiloxane nanostructures (silicone nanofilaments) synthesized | Superhydrophobicity | [109] |
| 49 | Composite based on polysiloxane and silver-coated copper flakes | Functional polymers | [110] |
| 50 | Polyhydroxyurethane bearing silicone backbone prepared by polyaddition | Functional materials | [111] |
| 51 | α,ω-bis(3-(1-methoxy-2-hydroxypropoxy) propyl) polydimethylsiloxane and α-N,N-dihydroxyethylaminopropyl-ω-butylpolydimethylsiloxane used to prepare block and graft waterborne polyurethane–polysiloxane copolymer dispersions | For graft waterborne polyurethane–polysiloxane film formation | [112] |
| 52 | Cu (I)-catalyzed azide-alkyne 1,3-dipolar cycloaddition reaction employed siloxane-based amphiphilic polymers containing alkyne moieties using click chemistry | Nanoemulsions useful as drug carriers | [113] |
| 53 | Synthesis of hydrophobic silica nanoparticles using a mixture of poly(dimethylsiloxane) and diethylcarbonate | Nanofillers of various polymeric systems for coatings | [114] |
| 54 | Water-repellent outer layer is applied using poly(dimethylsiloxane) (PDMS)–isocyanate | Food- and nonfood-contact packaging applications | [115] |
| 55 | Cross-linker polymethyl(ketoxime)siloxane (PMKS) with dense pendant reactive groups based on PMHS | As a cross-linker | [116] |

PDMS shows many properties that are not common in organic polymers such as hydrophobicity, high flexibility, low viscosity, and good thermal stability. These properties have made these polymers efficient to be used in various applications such as high-temperature insulation, antifoam applications, biotransplants, drug-delivery systems, and flexible elastomers, personal products. Other than these, many of the polymers are fire-retardant in contrast to many organic polymers that are flammable. An organosilicon polymer has aided to fill the gap between mineral silicates on one side and other side organic polymers.

The glass transition temperatures ($T_g$) of a polymer is to determine the entangle freedom of the backbone chain of a polymer. $T_g$ value above the freedom of motion on the backbone chain of a polymer segments and on the other hand below its $T_g$ shows the glassy state. Generally, an elastomeric polymer shows low $T_g$. For example, the $T_g$ value of natural rubber is −72 °C, and that of polyisobutylene is −70 °C. Thus, poly(dimethylsiloxane) shows lower $T_g$ value as compared to known elastomers. After studying the $T_g$ value of poly(dimethylsiloxane) and iso-electronic polyphosphazene $[NP(CH_3)_2]_n$, the polyphosphazene shows a $T_g$ value of −46 °C. The response of poly(diphenylsiloxane) [PDMS] changed by replacement of methyl substituent. PDMS shows the $T_g$ value of +49 °C; after the replacement of one phenyl group with a methyl group, the $T_g$ value increases. The same effect is observed in alternating or irregular copolymers. On the other hand, very curiously, methyl group replaced by tiny substituent "H" decreases the $T_g$ value up to 139 °C for $[MeHSiO]_n$. In PDMS, the bond length of Si–O is ~1.64 Å, and the distance of Si–C is in the range of 1.87–1.90 Å. Generally, organic polymers show that the bond length of C–C is 1.54 Å, and the same tetrahedral ($T_d$) bond length is observed in silicon and also shows the coil-type morphology [117]. Finally, inter-molecular coordination in PDMS is almost negligible, which shows the trans–syn-type structure [118]. PCSNs show remarkable properties in aerospace, automotive, microelectronics, and construction industries due to their special quality, that is, high-temperature resistance [119–125].

## 1.4 Applications

Polysilicon thin film that is used in the transistor shows the enhanced field effect mobility [126]. The porous polysilicon material is employed for humidity sensing as well as for biosensing purpose [127, 128] and also it is a versatile material for micro-systems. Poly Si-TFT material showed memory storage capacity polysilicon films developed by catalytic CVD, and then used in a solar cell panel [129, 130].

## 1.5 Conclusion

Polysiloxanes belong to the group of polymeric organosilicon compounds that are optically clear, inert, and nonflammable, with very special applications in medical and engineering fields. These can be synthesized by anionic ring-opening polymerization of hexamethylcyclotrisiloxane and it can also form a block copolymer and many other macromolecular architectures with low polydispersity index and homogeneity.

# References

[1]    Riedel, R., Mera, G., Hauser, R., & Klonczynski, A. Silicon-based polymer-derived ceramics: Synthesis properties and applications a review, J. Ceram. Soc. Jpn., 2006, 114, 425–444.

[2]    Colombo, P., Mera, G., Riedel, R., & Sorarˇu, G. D. Polymer-derived ceramics: 40 years of research and innovation in advanced ceramics, J. Am. Ceram. Soc., 2010, 93, 1805–1837.

[3]    Barroso, G. S., Krenkel, W., & Motz, G. Low thermal conductivity coating system for application up to 1000 °C by simple PDC processing with active and passive fillers, J. Eur. Ceram. Soc., 2015, 35(12), 3339–3348.

[4]    G¨unthner, M., Sch¨utz, A., Glatzel, U., Wang, K., Bordia, R. K., Greißl, O., Krenkel, W., & Motz, G. High performance environmental. barrier coatings, part 1: Passive filler loaded SICN system for steel, J. Eur. Ceram. Soc., 2011, 31, 3003–3010.

[5]    Prasad, R. M., Iwamoto, Y., Riedel, R., & Gurlo, A. Multilayer amorphous-Si-B-C-N/γ-Al2O3/α-Al2O3 membranes for hydrogen purification, Adv. Eng. Mater., 2010, 12(6), 522–528.

[6]    Konegger, T., Williams, L. F., & Bordia, R. K. Planar, polysilazane-derived porous ceramic supports for membrane and catalysis applications, J. Am. Ceram. Soc., 2015, 98(10), 3047–3053.

[7]    Yan, X., Gottardo, L., Bernard, S., Dibandjo, P., Brioude, A., Moutaabbid, H., & Miele, P. Ordered mesoporous silicoboroncarbonitride materials via preceramic polymer nano casting, Chem. Mater., 2008, 20(20), 6325–6334.

[8]    Majoulet, O., Salameh, C., Schuster, M. E., Demirci, U. B., Sugahara, Y., Bernard, S, & Miele, P Preparation, characterization, and surface modification of periodic mesoporous silicon–aluminum–carbon–nitrogen frameworks, Chem. Mater., 2013, 25(20), 3957–3970.

[9]    Marceaux, S., Bressy, C., Perrin, F., Martin, C., & Margaillan, A. Development of polyorganosilazane-silicone marine coatings, Prog. Org. Coat., 2014, 77, 1919–1928.

[10]   Vu, C., Osterod, F., Brand, S., & Ryan, K. Silicon and nitrogen Eur. Coat. J., 2008, 38, 1–11.

[11]   G¨unthner, M., Wang, K., Bordia, R. K., & Motz, G. Conversion behavior and resulting mechanical properties of polysilazane-based coatings, J. Eur. Ceram. Soc., 2012, 32(9), 1883–1892.

[12]   Herzog, A., Thünemann, M., Vogt, U., & Beffort, O. Novel application of ceramic precursors for the fabrication of composites, J. Eur. Ceram. Soc., 2005, 25(2–3), 187–192.

[13]   Birot, M., Pillot, J. P., & Dunogues, J. Comprehensive chemistry of polycarbosilanes, polysilazanes, and polycarbosilazanes as precursors of ceramics, Chem. Rev., 1995, 95, 1443–1477.

[14]   Yajima, S., Omori, M., & Hayashi, J. Simple synthesis of continuous SiC fiber with high tensile strength, Chem. Lett., 1976, 5, 551–554.

[15]   Johnson, D. W., Evans, A. G., & Goettler, R. W. Ceramic Fibers and Coatings: Advanced Materials for the Twenty-First Century MI. Washington D C, National Academy Press, 1998, 1–49.

[16]   Le, P., Monthioux, M., & Oberljn, A.,Understanding Nicalon fiber, J. Euro. Ceram. Soc., 1993, 11, 95–103.

[17]   Brutsch, R. Chemical vapor deposition of silicon carbide and its applications. Thin Solid Films, 1985, 126, 313–318.

[18]   Golecki, I., Reidinger, F., & Marti, J. Single-crystalline, epitaxial cubic SiC films grown on (100) Si at 750°C by chemical vapor deposition, Appl. Phys. Lett., 1992, 60, 1703–1705.

[19]   Colombo, P., Martucci, A., Fogato, O., & Villoresi, P. Silicon carbide films by laser pyrolysis of polycarbosilane. J. Am. Ceram. Soc., 2001, 84, 224–226.

[20]   Feng, Z. C. Second-order Raman scattering of cubic silicon carbide. In Proceedings of the XIXth International Conference on Raman Spectroscopy, ed. P. M. Fredericks, R. L. Frost, and L. Rintoul. CSIRO Publishing, Gold Coast, Queensland, Australia, 2004, 242–243.

[21] Katharria, Y. S., Kumar, S., Prakash, R., Choudhary, R.J., Singh, F., Phase, D. M., & Kanjilal, D. Characterizations of pulsed laser deposited SiC thin films, J. Non-Cryst. Solids., 2007, 353, 4660–4665.

[22] Sha, Z.D., Wu, X. M., & Zhuge, L. J. Initial study on the structure and photoluminescence properties of SiC films doped with Al, Appl. Surf. Sci., 2006, 252, 4340–4344.

[23] Polychroniadis, E., Syvajarvi, M., Yakimova, R., & Stoemenos., J. Microstructural characterization of very thick freestanding 3C.SiC wafers, J. Cryst. Growth, 2004, 263, 68–75.

[24] Agarwal, P., Mishra, P. K., & Srivastava, P. Statistical optimization of the electrospinning process for chitosan/polylactide nanofabrication using response surface methodology, J. Mater. Sci., 2012, 47, 4262–4269.

[25] Sigmund, W., Yuh, J., Park, H., Maneeratana, V., Pyrgiotakis, G., Daga, A., Taylor, J., & Nino, J. C. Processing and structure relationships in electrospinning of ceramic fiber systems. J. Am. Ceram. Soc., 2006, 89, 395–407.

[26] Greiner, A., & Wendorff, J. H. Electrospinning: A fascinating method for the preparation of ultrathin fibers, Angew. Chem. Int. Ed., 2007, 46, 5670–5703.

[27] Tang, M., Su, Z., Wang, Z., Zhang, L., & Chen, L. Irradiation pre-curing plus oxidation curing for rapid preparation of silicon carbide fibers, J. Mater. Sci., 2009, 44, 3905–3908.

[28] Narisawa, M., Idesaki, A., Kitano, S., Okamura, K., Sugimoto, M., Seguchi, T., & Itoh, M. Use of the blended precursor of poly(vinylsilane) in polycarbosilane for silicon carbide fiber synthesis with radiation curing, J. Am. Ceram. Soc., 1999, 82, 1045–1051.

[29] Mark, J. E., West, R., & Allcock, H.R. Inorganic polymers, Prentice-Hall, Englewood Cliffs, 1992.

[30] Cypryk, M., & Apeloig, Y. Mechanism of the acid-catalyzed Si-O bond cleavage in siloxanes and siloxanols. A theoretical study, Organometallics, 2002, 21, 2165–2175.

[31] Yajima, S., Hasegawa, Y., & Hayashi, J. Synthesis of continuous silicon carbide fiber with high tensile strength and high young's modulus. J. Mater. Sci., 1978, 13, 2569–2576.

[32] Wang, Y. D., Feng, C. X., & Song, Y. C. Study on the process of continuous SiC fibers, Aerospace Mater. Tech., 1997, 2, 21–25.

[33] Okamura, K., & Seguchi, T. Application of radiation curing in the preparation of polycarbosilane-derived SiC fibers, J. Inorg. Organomet. Polym., 1992, 2, 171–179.

[34] Youngblood, G. E., Jones, R. H., Kohyama, A., & Snead, L. L. Radiation response of SiC-based fibers,J. Nucl. Mater., 1998, 258–263, 1551–1556.

[35] Seyferth, D., Wiseman, G.H., & Homme, C. P. High-yield synthesis of $Si_3N_4$/SiC ceramic materials by pyrolysis of a novel polyorganosilazane, J. Am. Ceram. Soc., 1983, 66, 13–17.

[36] Xie, Z. M., & Li, G. L. Advances in researches of polysilazane, Polym Bull., 1995, 12, 138–144.

[37] Wang, C., Song, N., Ni, L., & Ba, C. Synthesis, thermal properties, and ceramization of a novel ethynylaniline-terminated polysilazane, High Per. Polym., 2016, 28, 359–367.

[38] Seo, Y., Cho, S., Kim, S., Choi, S., & Kim, H. Synthesis of refractive index tunable silazane networks for transparent glass fiber reinforced composite. Ceramics Int., 2017, 43(10), 7895–7900.

[39] Lei, Y.P., & Song, Y.C. Boron nitride by pyrolysis of the melt-processable poly [tris (methylamino) borane]: Structure, composition and oxidation resistance Ceramics Int., 2012, 38, 271–276.

[40] Yajima, S., Hayashi, J., & Omori, M Continuous silicon carbide fiber of high tensile strength, Chem. Lett., 1975, 4, 931–934.

[41] Kusari, U., Bao, Z., Cai, Y, Ahmad, G., Sandhge, K.H., & Sneddon, L.G. Formation of nanostructured, nanocrystalline boron nitride microparticles with diatom-derived 3-D shapes, Chem. Commun., 2007, 11, 1177–1179.

[42] Yan, M., Song, W., & Chen, Z.H. In situ growth of a carbon interphase between carbon fibres and a polycarbosilane-derived silicon carbide matrix, Carbon, 2011, 49, 2869–2872.

[43] Tang, Y., Wang, J.;, Li, X, Xie, Z., Wang, H., Li, W., & Wang, X. Polymer-derived SiBN fiber for high temperature structural/functional applications, Chem. Eur. J., 2010, 16,6458–6462.

[44] Lee, J, Butt, D P, Baney, R H, Bowers, C.R., & Tulenko, J.S. Synthesis and pyrolysis of novel polysilazane to SiBCN ceramic, J. Non-Cryst. Solids, 2010, 351, 2995–3005.

[45] Zhang, C., Han, K., Liu, Y., Mou, S., Chang, X., Zhang, H., Ni, J., & Yu, M. A novel high yield polyborosilazane precursor for SiBNC ceramic fibers, Ceramic Int., 2017, 43(13), 10576–80.

[46] Furtat, P., Lenz-Leite, M., Ionescu, E., Machado, R. A. F., & Motz, G. Synthesis of fluorine-modified polysilazanes via Si– H bond activation and their application as protective hydrophobic coatings, J. Mat. Chem. A, 2017, 5, 25509–25521.

[47] Schwark, J.M.W. Isocyanate- and isothiocyanate-modified polysilazane ceramic precursors, 1996. EP Patent 0,442,013, August 21.

[48] Bauer, M., Decker, D., Richter, F., & Gwiazda, M. DE 2010.Patent 102,009,013,410,

[49] Steffen, S., Bauer, M., Decker, D, & Richter, F. Fire-retardant hybrid thermosetting resins from unsaturated polyesters and polysilazanes, J. Appl. Polym. Sci., 2014, 131(12),40375 (1–7).

[50] Jones, R. G. Silicon-containing polymers. The Royal Society of Chemistry; Cambridge 1995.

[51] Zamora, M., Bruna, S., Alonso, B., & Cuadrado, I. Polysiloxanes bearing pendant redox-active dendritic wedges containing ferrocenyl and (h6-aril) tricarbonylchromium moieties, Macromolecules, 2011, 44, 7994–8007.

[52] Unno, M., Tanaka, R., Tanaka, S., Takeuchi, T., Kyushin, S., & Matsumoto, H. Oligocyclic ladder polysiloxanes: Alternative synthesis by oxidation, Organometallics, 2005, 24, 765–768.

[53] Chvalovsky, V., Haiduc, I., Sowerby, D. B. (eds) The chemistry of inorganic homo-and heterocycles. Academic Press, London, 1987, 287–348

[54] Babu, N.G., Christopher, S. S., & Newmark, A. R. Poly(dimethylsiloxane-co-diphenylsiloxanes): Synthesis, characterization, and sequence analysis, Prentice Hall, Englewood Cliffs, Macromolecules, 1987, 20(11),2654–2659.

[55] Chojnowski, J., Cypryk, M., Fortuniak, W., Cibiorekand, S. A. M., & Rozga-Wijas, K. Synthesis of branched polysiloxanes with controlled branching and functionalization by anionic ring-opening polymerization, Macromolecules, 2003, 36(11), 3890–3097.

[56] Brochon, C., Mingotaud, F. A., Schappacher, M., & Soum, A. Equilibrium anionic ring-opening polymerization of a six-membered cyclosiloxazane, Macromolecules, 2007, 40(10), 3547–3553.

[57] Jiang, B., Zhang, K., Zhang, T., Xu, Z., & Huang, Y. Investigation of reactivity and biocompatibility poly-p-phenylenebenzobisoxazole fiber grafted hyperbranched polysiloxane, Comp. B Eng., 2017, 121, 1–8.

[58] Wu, C., Yu, J., Li, Q., & Liu, Y. High molecular weight cyclic polysiloxanes from organocatalytic zwitterionic polymerization of constrained spirocyclosiloxanes. Polym. Chem., 2017, 8, 7301–7306.

[59] Ning, P. Y., Tang, Y. M., Jiang, Y. C., Mark, E. J., & Roth, C. W. Particle sizes of reinforcing silica precipitated into elastomeric networks, J. Appl. Polym. Sci., 1984, 29(10), 3209–3212.

[60] Bokobza, L., & Diop, L. A. Reinforcement of poly(dimethylsiloxane) by sol-gel in situ generated silica and titania particles, Exp. Polym. Lett., 2010, 4(6), 355–363.

[61] Wen, J., & Mark, E. J. Synthesis, structure, and properties of poly(dimethylsiloxane) networks reinforced by in situ-precipitated silica-titania, silica–Zirconia, and silica-alumina mixed oxides, J. Appl. Polym. Sci., 1995, 58(7), 1135–1145.

[62] Chen, Y., Yu, P., Feng, C., Han, Q., & Zhang, Q. Synthesis of polysiloxane with quaternized N-halamine moieties for antibacterial coating of polypropylene via supercritical impregnation technique, Appl. Surf. Sci., 2017, 419, 683–691.

[63] Loh, C. T., Ng, M. C., Kumar, N. R., Ismail, H., & Ahmad, Z. Improvement of thermal ageing and transparency of methacrylate-based poly(siloxane–Silsesquioxane) for optoelectronic application, J. Appl. Polym. Sci., 2017, 134 (37), 45285 (1–11).

[64] Bharwal, K. A., Nguyen, A. N., Iojoiu, C., & Henrist, C. New polysiloxane bearing imidazolium iodide side chain as electrolyte for photoelectrochemical cell, Solid State Ionics, 2017, 307, 6–13.

[65] Ma, Y., He, L., Jia, M., Zhao, L., Zuo, Y., & Hu, P. Cage and linear structured polysiloxane/epoxy hybrids for coatings: Surface property and film permeability, J. Colloid Interface Sci., 2017, 500, 349–357.

[66] Donmez, B. K., Gencten, M., & Sahin, Y. A novel polysiloxane-based polymer as a gel agent for gel–VRLA batteries, Ionics, 2017, 23(8), 2077–2089.

[67] Zhang, G., Jiang, S., Gao, Y., & Sun, F. Design of green hydrophilic polysiloxane–Polyether-modified macromolecular photoinitiators with ionic liquid character J. Mat. Sci., 2017, 52(16), 9931–9945.

[68] Shim, J., Kim, L., Kim, H. J., Jeong, D., Lee, H.; J, & Lee, C. J. All-solid-state lithium metal battery with solid polymer electrolytes based on polysiloxane crosslinked by modified natural gallic acid. Polymer (United Kingdom), 2017, 122, 222–231.

[69] Jiang, B., Zhang, K., Zhang, T., Xu, Z., & Huang, Y. Investigation of reactivity and biocompatibility poly-*p*-phenylene benzobisoxazole fiber grafted hyperbranched polysiloxane Composite Part B Eng., 2017, 121, 1–7.

[70] Liu, C., Liu, M. H., Huang, Y. R., & Chen, C. L. Molecular dynamics simulation of wave-absorbing materials based on polysiloxane, Mater. Chem. Phys., 2017,195, 10–18.

[71] Stular, D., Vasiljevic, J., Colovic, M., Mihelcic, M., Medved, J., Kovac, J., Jerman, I., Simoncic, B., & Tomsic, B. Combining polyNiPAAm/chitosan microgel and bio-barrier polysiloxane matrix to create smart cotton fabric with responsive moisture management and antibacterial properties: Influence of the application process, J. Sol-Gel Sci. Tech., 2017, 83(1), 19–34.

[72] Li, S., Li, H., Li, Z., Zhou, H., Guo, Y., Chen, F., & Zhao, T. Polysiloxane modified phenolic resin with co-continuous structure Polymer United Kingdom, 2017, 120, 217–219.

[73] Chen, Y., Zhang, O., Han, O., Mi, Y., Sun, S., Feng, C., Xiao, H., Yu, P., & Yang, C. Synthesis of polysiloxane with 5,5-dimethylhydantoin-based *N*-halamine pendants for biocidal functionalization of polyethylene by supercritical impregnation J. Appl. Polym. Sci., 2017, 134, 107.

[74] Wang, G., Zhu, Y., Zhai, Y., Wang, W., Du, Z., & Qin, J. Polyglycerol modified polysiloxane surfactants: Their adsorption and aggregation behavior in aqueous solution, J. Ind. Eng. Chem., 2017, 47,121–127.

[75] Li, Z., Wu, C., Liu, L., Li, M., Yang, X., Hao, C., Chen, O., Hu, Z., Luo, M., Lai, G., & Luh, T.Y. Chemoselective synthesis of macrocyclic poly(methylvinylsiloxane)s via metathetical ring-expanding polymerization of oligomeric cyclosiloxanes by sequential (MeVinylSi-O-) insertion reactions, Polym. Chem., 2017, 8(9), 1573–1578.

[76] Chang, C. J., Lai, C. F., Chiou, W. Y., Reddy, P. M., & Su, M. J. Effects of hydrosilyl monomers on the performance of polysiloxane encapsulant/phosphor blend based hybrid white-light-emitting diodes J. Appl. Polym. Sci.,2017, 134, 44524.

[77] Wang, X., Sun, S., Wang, H., & Guo, X. Synthesis and characterization of polysiloxane grafted polyamide-amine surfactants, Surfactants Deterg., 2017, 20(2), 521–528.

[78] Dunki, J., Cuervo-Reyes, E., & Opris, D. M. A facile synthetic strategy to polysiloxanes containing sulfonyl side groups with high dielectric permittivity, Polym. Chem., 2017, 8(4), 715–724.

[79] Sun, Z., Liu, B., Huang, S., Wu, J., & Zhang, Q. Facile fabrication of superhydrophobic coating based on polysiloxane emulsion, Prog. Org. Coat., 2017, 102, 131–137.

[80] Byczyński, Ł., Dutkiewicz, M., & Maciejewski, H. Thermal and surface properties of hybrid materials obtained from epoxy-functional urethane and siloxane. Polym. Bull. 2016, 73, 5, 1247–1265.

[81]  Peng, Z., Yi, D., Xue, Z., Li, H., Li, Q., & Hu, Y. Synthesis of dendritic polyethylene brushes with polysiloxane as the main chain, Macromol. Chem. Phy., 2017, 218, 700143.

[82]  Dong, J., Yang, C., Cheng, Y., Wu, T., Zhao, X., & Zhang, O. Facile method for fabricating low dielectric constant polyimide fibers with hyperbranched polysiloxane, J. Mat. Chem., 2017, 5 (11), 2818–2825.

[83]  Lee, E., Jung, J., Choi, A., Bulliard, X., Kim, J. H., Yun, Y., Kim, J., Park, J., Lee, S., & Kang, Y. Dually crosslinkable SiO$_2$@polysiloxane core-shell nanoparticles for flexible gate dielectric insulators, RSC Adv., 2017, 7(29), 17841–17847.

[84]  Dundua, A., Franzka, S., & Ulbricht, M. Improved antifouling properties of polydimethylsiloxane films via formation of polysiloxane/polyzwitterion interpenetrating networks, Macromol. Rapid Comm., 2016, 37(24), 2030–2036.

[85]  Li, Q., Yang, Z., Ren, Z., & Yan, S. Polysiloxane- modified tetraphenylethene: Synthesis, AIE properties, and sensor for detecting explosives, Macromol. Rapid Comm., 2016, 37(21), 1772–1779.

[86]  Wang, C., & An, O. Synthesis, film morphology and performance of polysiloxane having wax groups and polyether amino groups, Polym. Mat. Sci. Eng., 2016, 32, 36–41.

[87]  Boaretto, N., Joost, C., Sevfried, M., Vezzu, K., & Noto, V. D. Conductivity and properties of polysiloxane-polyether cluster-LiTFSI networks as hybrid polymer electrolytes, J. Power Sources, 2016, 325, 427–437.

[88]  Tyszczuk-Rotko, K., Sadok, I., & Barczak, M. Thiol-functionalized polysiloxanes modified by lead nanoparticles: Synthesis, characterization and application for determination of trace concentrations of mercury (II), Micropor. Mesopor. Mat., 2016, 230, 109–117.

[89]  Lakiza, N.V., & Neudachina, L.K. Russian, synthesis and physicochemical properties of polysiloxane functionalized with amino acetic acid groups, Russ. J. Phys. Chem. A, 2016, 90(7), 1450–1455.

[90]  Zuo, Y., Gou, Z., Zhang, C., & Feng, S. Polysiloxane-based autonomic self-healing elastomers obtained through dynamic boronic ester bonds prepared by thiol-ene "click" chemistry, Macromol. Comm., 2016, 37(15), 1052–1059.

[91]  Ahmed, M.A., Abu Shaweesh, A. A., Ashgar, N.M., El-Nahhal, I.M., Chehimi, M. M., & Babonneau, F. Synthesis and characterization of immobilized-polysiloxane monoamine-thiol triacetic acid and its diamine and triamine derivatives, J. Sol-Gel Sci. Tech., 2016, 78(3), 660–672.

[92]  Ichikawa, T., Wako, T., & Nemoto, N. Synthesis of ionic liquid based on polysiloxane with quaternized imidazolium moiety, Polymer Bulletin, 2016, 73(5), 1361–1371.

[93]  Fei, H-F., Han, X., Liu, B., Gao, X., Wang, Q., Zhang, Z., & Xie, Z. Synthesis of gradient copolysiloxanes by simultaneous copolymerization of cyclotrisiloxanes and mechanism for kinetics inverse between anionic and cationic ring-opening polymerization, J. Polym. Sci., A; Chem., 2016, 54(6), 835–843.

[94]  Stochmal, E., Strzezik, J., & Krowiak, A. Preparation and characterization of polysiloxane networks containing metallic platinum particles, J. Appl. Polym. Sci., 2016, 133, 43096. (1–14)

[95]  Chen, N., Reeja-Jayan, B., Liu, A., Lau, J., Dunn, B., & Gleason, K. K. iCVD cyclic polysiloxane and polysilazane as nanoscale thin-film electrolyte: Synthesis and properties, Macromol. Rapid Comm., 2016, 37(5),446–452.

[96]  Ichikawa, T., Wako, T., & Nemoto, N. Synthesis of polysiloxane-based quaternized imidazolium salts with a hydroxy group at the end of alkyl groups, React. Funct. Polym., 2016, 99, 1–8.

[97]  Yao, W., Gao, Y., Yuan, X., He, B., Yu, H., Zhang, L., Shen, Z., He, W., Yang, Z., Yang, H., & Yang, D. Synthesis and self-assembly behaviours of side-chain smectic thiol–Ene polymers based on the polysiloxane backbone, J. Mat. Chem., 2016, 4(7), 1425–1440.

[98] Zhang, L., Chen, S., Zhao, H., Shen, Z., Chen, X., Fan, X., & Zhou, Q. Synthesis and properties of a series of mesogen-jacketed liquid crystalline polymers with polysiloxane backbones, Macromolecules, 2010, 43(14), 6024–6032.

[99] Sun, D., Zhou, X., Liu, J., Sun, X., Li, H., Ren, Z., Ma, D., Bryce, M.R., & Yan, S. Solution-processed blue/deep blue and white phosphorescent organic light-emitting diodes (PhOLEDs) hosted by a polysiloxane derivative with pendant mCP (1,3-bis(9-carbazolyl) benzene), ACS Appl. Mat. Interf., 2015, 7(51), 27989–27998.

[100] Yu, B., Kang, S. Y., Akthakul, A., Ramadurai, N., Pilkenton, M., Patel, A., Nashat, A., Anderson, D. G., Sakamoto, F.H., Gilchrest, B.A., Anderson, R.R., & Langer, R. An elastic second skin. Nat. Mat., 2016, 15(8), 911–918.

[101] Xu, T., Liu, H., Shang, S.B., Song, Q.Z., & Yang, C. Synthesis and properties of hydrogenated rosin modified vinyl polysiloxane. Chem. Ind. forest Prod., 2015, 35(6), 83–88.

[102] Zhao, J., An, Q. F, Li, X. Q., Huang, X. L., & Xu, X. A comb like polysiloxane with pendant quaternary ammonium polyether groups: Its synthesis, physical properties and antibacterial performance J. Polym. Res., 2015, 22(9), 174.

[103] Wang, H., Liu, W., Yan, Z., Tan, J., & Xia-Hou, G. Synthesis and characterization of UV-curable acrylate films modified by functional methacrylate terminated polysiloxane hybrid oligomers, RSC Adv., 2015, 5(100), 81838–81846.

[104] Liang, Y., Yuang, F., Lu, Y., Lu, H., & Feng, S. The synthesis and structure analysis of a novel polysiloxane-lanthanide hybrid material, Rus. J. Chem. A 2015, 89(9), 1619–1624.

[105] Ma, S., Li, X., Bai, L., Lan, X., Zhou, N., & Meng, F. Synthesis and characterization of imidazolium-based polymerized ionic liquid crystals containing cholesteryl mesogens, Coll. Polym. Sci., 2015, 293(8), 2257–2268.

[106] EL-Sukkary, M. M. A., Ismail, D. A., Rayes, S. M. E., & Saad, M. A. Synthesis, characterization and surface properties of amino-glycopolysiloxane. J. Ind. Eng. Chem., 2014, 20(5), 3342–3348.

[107] Sun, D., Yang, Z., Sun, X., Li, H., Ren, Z., Liu, J., Ma, D., & Yan, S. Synthesis of triphenylamine based polysiloxane as a blue phosphorescent host, Polym. Chem., 2014, 5(17), 5046–5052.

[108] An, Q., Zhao, J., Li, X., Wei, Y., & Qin, W. Synthesis of dimethyldodecyl quaternary ammonium polyether polysiloxane and its film morphology and performance on fabrics. Appl. Sci., 2014. 131(16), 40612(1–8).

[109] Korhonen, J.T., Huhtamaki, T., Verho, T., & Ras, R.H.A. Hollow polysiloxane nanostructures based on pressure-induced film expansion, Surface Innovations, 2014, 2(1), 116–126.

[110] Xia, Q M., Liu, K. H., Chen, S. L., Gao, Y. Q., Luo, Y. F., & Liu, L. Acta PolymericaSinica,2014,1, 1.

[111] Ochiai, B., Kojima, H., & Endo, T. Synthesis and properties of polyhydroxyurethane bearing silicone backbone, J. Polym. Sci. A: Polym. Chem., 2014, 52(8), 1113–1118.

[112] Yu, Y., & Wang, J. Synthesis and properties of block and graft waterborne polyurethane modified with α,ω-bis(3-(1-methoxy-2-hydroxypropoxy)propyl)polydimethylsiloxane and α-N, N-dihydroxyethylaminopropyl-ω-butylpolydimethylsiloxane, Polym. Eng. Sci., 2014, 54(4), 805–811.

[113] Kihara, Y., Ichikawa, T., Abe, S., Nemoto, N., Ishihara, T., Hirano, N., & Haruki, M. Synthesis of alkyne-functionalized amphiphilic polysiloxane polymers and formation of nanoemulsions conjugated with bioactive molecules by click reactions, Polym. J. 2014, 46(3), 175–183.

[114] Protsak, I., Pakhlov, E., Tertykh, V., Le, Z. C., & Dong, W. A new route for the preparation of hydrophobic silica nanoparticles using a mixture of poly(dimethylsiloxane) and diethyl carbonate, Polymers, 2018, 10(2), 116 (1–13).

[115] Li., Zhao, & Rabnawaz, M. Fabrication of food-safe water-resistant paper coatings using a melamine primer and polysiloxane outer layer, ACS Omega, 2018, 3(9), 11909–11916.

[116] Zhan, Xi, Caib, X., & Zhang, J. A novel crosslinking agent of polymethyl(ketoxime) siloxane for room temperature vulcanized silicone rubbers: Synthesis, properties, and thermal stability, RSC Adv., 2018, 8(23), 12517–12525.

[117] ClarsonIn, J. S., Clarson, J. S., & Semlyen, A. J. Depolymerisation, degradation and thermal properties of siloxane polymers, (eds) Siloxane polymers. Prentice-Hall, Englewood Cliffs, 1993, 216–244.

[118] Dvornic, R. P., & Lenz, W. R. Exactly alternating silarylene-siloxane polymers. 9. Relationships between polymer structure and glass transition temperature, Macromolecules, 1992, 25(14), 3769–3778.

[119] Lewinsohn, C. A., Henager, C. H., Youngblood, G. E., Jones, R.H., Lara-Curzio, E., & Scholz, R. Failure mechanisms in continuous-fiber ceramic composites in fusion energy environments. J. Nucl. Mater., 2001, 289(1–2), 10–15.

[120] Homrighausen, C. L., & Keller, T. M. High-temperature elastomers from silarylene-siloxane-diacetylene linear polymers, J Polym. Sci. A, 2002, 40(1), 88–94.

[121] Wang, R., Liu, W., & Fang, L. Synthesis, characterization, and properties of novel phenylene-silazane-acetylene polymers, Polymer, 2010, 51(25), 5970–5976.

[122] Fang, Y. H., Huang, M. H., Yu, Z. J., Xia, H., Chen, L., Zhang, Y., & Zhang, L. Synthesis, characterization, and pyrolytic conversion of a novel liquid polycarbosilane, J. Am. Ceram. Soc., 2008, 91(10), 3298–3302.

[123] Birot, M., Pillot, J. P., & Dunogues, J. Ex-cellulose carbon fibers with improved mechanical properties. J. Mater. Sci., 2006, 7, 1959–1964.

[124] Itoh, M., Inoue, K., & Iwata, K. New highly heat-resistant polymers containing silicon: Poly (silyeneethynylenephenyleneethynylene)s, Macromolecules, 1997, 30, 697–701.

[125] Craig, L. H., & Teddy, M. K. Synthesis and characterization of a silarylene-siloxane-diacetylene polymer and its conversion to a thermosetting plastic, Polymer, 2002, 43, 2619–2623.

[126] Kuriyama, H., Kiyama, S., Noguchi, S., Kuwahara, Takashi, Ishida, Satoshi, N., Tomoyuki, S., Keiichi, Iwata., Hiroshi, K., & Hiroshi, O. Enlargement of poly-Si film grain size by excimer laser annealing and its application to high-performance poly-Si thin film transistor, Jpn. J. Appl. Phys., 1991, 30(125), 3700.

[127] Islam, T., Pramanik, C., & Saha, H., Modeling, simulation and temperature compensation of porous polysilicon capacitive humidity sensor using ANN technique, Microelectron Reliab., 2005, 45(3–4), 697–703.

[128] Desai, T. A., Hansford, D. J., Leoni, L., Essenpreis, M., & Ferrari, M. Nanoporous anti-fouling silicon membranes for biosensor applications. Biosens. Bioelectron., 2000, 15(9–10), 453–462.

[129] Yin, H., Xianyu, W., Tikhonovsky, A., & Park, Y.S. Scalable 3-D fin-like poly-Si TFT, and its nonvolatile memory application, IEEE Trans. Electron Devices, 2008, 55(2), 578–584.

[130] Rath, J.K., Stannowski, B., van, V., Patrick, ATT, van, V., Marieke, K, & Ruud, EI Application of hot-wire chemical-vapor-deposited Si: H films in thin film transistors and solar cells, Thin Solid Films, 2001, 385(1–2), 320–329.

# 2 Tin- and germanium-based polymers: polystannanes and polygermanes

**Abstract:** In polystannanes, the mother chain consists of covalently bound tin atom, which shows a characteristic absorption at 375–410 (dialkylstannane) and 470–480 (diphenylstannane) entities in UV-vis spectra. These electronic absorption bands are attributed to σ-delocalization and σ–π-delocalization, respectively. Polygermanes are synthesized by using a developed form of Wurtz coupling and Grignard reactions, having a characteristic electronic band at 300–350 nm and narrow molecular distribution with a molecular weight of 10,000 daltons. In this chapter, several methods to synthesize both polygermanes and polystannanes have been discussed in detail.

**Keywords:** synthesis, polystannane, polygermane, coupling, electronics

## 2.1 Introduction

The main chain of polystannanes consists of covalently interconnected tin atoms. Because of the delocalization of electrons within the backbone of polymer (σ-delocalization), polystannanes are materials concerning their vast applications in many fields, namely chemical, optical, thermal, and electrical. High-molecular-weight (MW) polystannanes were developed by using dialkylstannanes ($H_2SnR_2$) with the Wilkinson's catalyst [$RhCl(PPh_3)_3$]. This route allows to obtain and isolate pure linear poly(dialkylstannane)s without cyclic oligomers; but on the other hand, has some substantial drawbacks. In particular, this method has so far not been suited to synthesize poly(diarylstannane)s. Hence, to create such materials a new synthetic route is required, for instance, reaction of dichlorodiorganostannanes in liquid ammonia.

Polystannane $(SnR_2)_n$ chain consists of tin atoms on the polymer backbone, which are covalently bonded. Polystannanes created interest due to stabilized covalently bounded metal atom backbone. Solution form of polystannanes can be easily decomposed photochemically. Poly(diorganostannane)s show good thermal stability above 200 °C. Liquid crystalline behavior property is observed in poly(dialkylstannane)s at room temperature.

In the last few years, copolymerization of polygermane and germane–silane generates more attention due to its special physical properties like semiconductivity, photoluminescence, photoconductivity, and third-order nonlinear effects [1, 2]. The synthesis of polygermane is carried out by reduction of organodichlorogermanes with the help of alkali metals [3]. After the study of this reaction method it is found that it is needed to maintain conditions that are very drastic and have polymodal MW distribution. For the synthesis of polysilane electroreduction of organodichlorosilane along with magnesium-based electrode, final yield showed a monomodal MW

https://doi.org/10.1515/9781501514609-003

distribution. A comparatively newer type of σ-conjugated photoconducting polymer consists of silicon–silicon (Si–Si)- and germanium–germanium (Ge–Ge)-type backbone, and organic substituents are polygermanes [4]. The delocalization of σ-electrons plays a vital role as a charge carrier in the polymer backbone chain via Ge and Si. The characteristics of polyethylene derivatives such as charge transporting and photoconducting are under investigation [5], and only one article has reported on poly(dibutylgennylene) and alkyl-substituted polygermane [6].

## 2.2 Synthetic methods

### 2.2.1 Polystannane

Synthesis of polystannanes can be classified into three different ways; the final yield is obtained in the form of cyclic pentamers and hexamers. These three routes are based on electrochemical reactions, Wurtz coupling, and catalytic dehydrogenation of tin dihydrides ($SnH_2$) [7]. Polystannanes were prepared using Wurtz reaction with sodium (Na) in organic solvents (Figure 2.1) [8, 9], and lightweight product and oligomeric byproducts are achieved. However, the straight-chain poly (dibutylstannane) was prepared by Wurtz reaction, resulting in high molar mass and number average molecular weight ($M_n$) up to $10^6$ g/mol [10]. Prepared materials are formed with accurate bimodal molar mass distribution, but are less reproducible with predominant oligomers [10, 11]. This same type of results is observed in

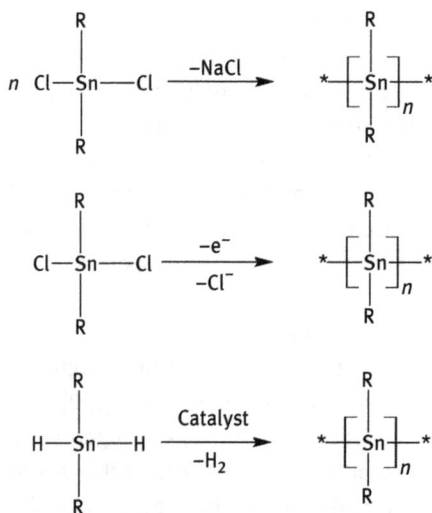

**Figure 2.1:** Preparation of polystannanes.

synthesized poly(diorganosilane) by Wurtz reaction with dichlorodiorganosilanes [12, 13]. This chain growth polymerization is initiated by metallic sodium surface in continuation with retardation step observed in the step of chain prolongation by the condensation reaction.

In the modified Wurtz reaction, dichlorodiorganostannanes ($R_2SnCl_2$) react with sodium (Na) in the presence of liquid ammonia [14]. Polymerization reaction can be carried out in two ways and is shown in Figure 2.2. Initially, four equivalents of sodium are added, which results in the formation of intermediate stannides [15, 16], followed by the addition of remaining dichlorodiorganostannane. In the two-step reaction method, two equivalents of sodium per dichlorodiorganostannane are obtained, as in the one-step reaction [17]. Following the second step reactions, dichlorodialkylstannanes are used as precursor compounds because of the replacement of alkyl groups in the intermediate stannides [15, 17]. On the contrary, under the applied conditions an exchange of phenyl groups is not observed.

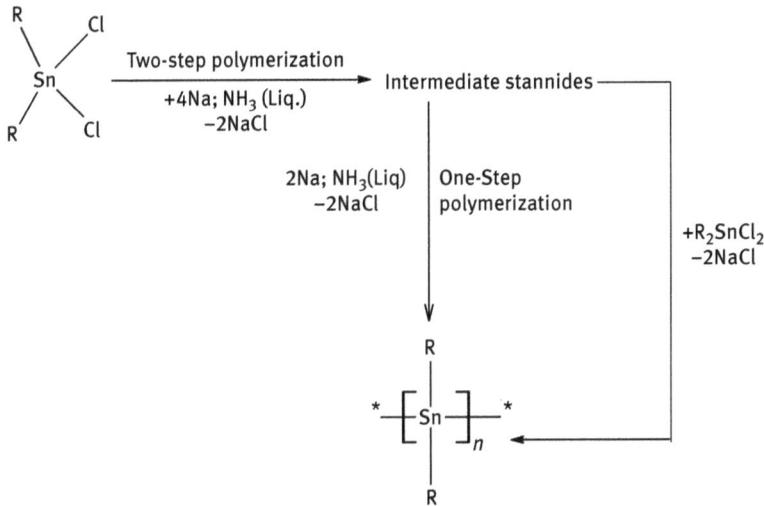

**Figure 2.2:** Synthesis part of polystannane.

Polystannane reacts with sodium in liquid ammonia medium, and homopolymers are obtained by using $R_2SnCl_2$ with R = butyl, phenyl, dodecyl, octyl, 3-methoxyphenyl. In addition, random copolymers are formulated by dialkylstannane and diphenylstannane groups [17, 18]. The synthesized polymers show weighted-average molar masses in the range of $1 \times 10^4$ to $2 \times 10^4$ g/mol. Figure 2.2 shows the preparation of polystannanes by electrochemical reactions, and dichlorodialkylstannanes are used as a starting material [19–21]. Figure 2.2 shows that the synthesis of polystannanes is done by catalytic dehydrocoupling using catalysts such as

platinum, rhodium, zirconium, or lanthanides [22]. In the preliminary stage, polymers synthesized by conventional techniques are generally unable to be extracted; As a result, polystannanes develop fewer amounts of products [23, 24].
Poly(dibutylstannane) is synthesized by dehydropolymerization of $Bu_2SnH_2$ in the
presence of heat, but it contains some amount of impurities [25]. The high percentage yield of poly(dialkylstannanes) is achieved for the polymerization of dialkylstannanes along with the catalyst [RhCl ($PPh_3$)$_3$] (Wilkinson's catalyst) and also
the formation of cyclic byproducts [26, 27]. Poly(dibutylstannane) is prepared by
the same catalytic method, which showed high MW range such as $1 \times 10^4$ to
$4 \times 10^4$ g/mol [28].

Copolymers and homopolymers of polystannanes are prepared in two steps via
the in situ method: stannides used as starting material and stannides obtained from
$R_2SnCl_2$ [17]. On the other hand, polystannanes can be achieved in a single step – the
mixture of two equivalent sodium (Na) per tin (Sn) atom along with $R_2SnCl_2/R'_2SnCl_2$
(Figure 2.3). Thus, poly(diphenylstannane) and poly(dibutylstannane) are prepared,
but both are insoluble in common solvents. Random $SnR_2/SnPh_2$ copolymers with
R = octyl or dodecyl, butyl, however, are partially soluble; weight average molar
masses counted ~10,000 g/mol.

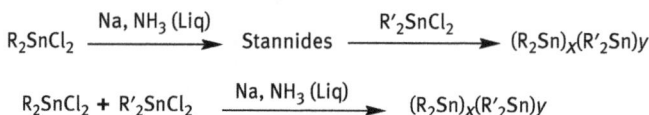

$R_2SnCl_2 \xrightarrow{\text{Na, NH}_3 \text{ (Liq)}} \text{Stannides} \xrightarrow{R'_2SnCl_2} (R_2Sn)_x(R'_2Sn)y$

$R_2SnCl_2 + R'_2SnCl_2 \xrightarrow{\text{Na, NH}_3 \text{ (Liq)}} (R_2Sn)_x(R'_2Sn)y$

**Figure 2.3:** Synthetic ways for polystannane homopolymers (R = R') and random copolymers in the
presence of Noah, liquid $NH_3$.

One gram of dialkylstannanes are mixed with a solution (30 mL) of [RhCl($PPh_3$)$_3$]
(4% mol/mol with respect to $R_2SnH_2$) in methylene chloride ($CH_2Cl_2$) under inert
argon gas. The reaction mixture is strictly kept away from light, stir the reaction
mixture for 2 h, and the solution is cooled to −78 °C for half an hour. The precipitated polymers are filtered, washed with small amounts of cold $CH_2Cl_2$ (−78 °C), and
then dried in vacuum.

*Poly[bis(3-methoxyphenyl)stannane]:* About 806.8 mg of dichlorobis (3-methox-
yphenyl)stannaneis mixed with 5 mL of tetrahydrofuran (THF), under nitrogen atmosphere in the presence of solution of Na (183.7 mg, 7.99 mmol) in liquid $NH_3$ (90 mL) at
−78 °C (Figure 2.4), and the reaction mixture is stirred for 30 min. Then, repeatedly
810.1 mg DCMS mixed in 5 mL of THF under exclusion of light. $NH_3$ was evaporated in
an $N_2$ stream, and THF was subsequently removed in vacuum at room temperature.
The remaining solids were washed with 20 mL mixture with ethanol/water to remove
NaCl.

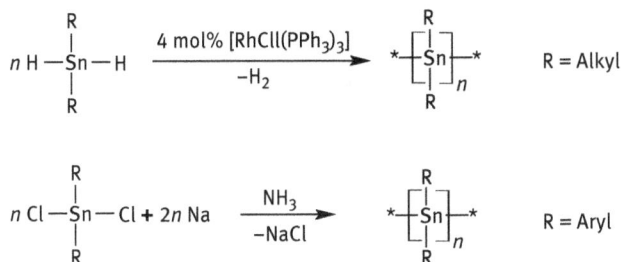

**Figure 2.4:** Synthesis pathway of polystannanes via catalyst [RhCl(PPh₃)₃] or by sodium in liquid ammonia.

## 2.2.2 Polygermane

Preparation of Ge(Me₃Si)₄ by using precursor-like GeCl₄, Me₃SiCl, and Li [29] quite enough butGeCl₄ is costly [30]. For the related synthesis of Ge(Me₃Ge)₄, a highly expensive Me₃GeCl compound is required. Recently, the same material is synthesized from Ge(Me)₄ by reaction with SnCl₄ [31]. Ge(Me)₄ is feasible from GeCl₄ and MeMgI [32]. After searching affordable precursor material conclude GeBr₄, which can be developed by germanium powder with bromine [33]. In contrast to the synthesis of Ge(Me₃Ge)₄reported [34], which used trimethylgermyllithium, finally develops (Me₃Ge) ₄ [35]. The reaction of tetrabromogermane and chlorotrimethyl germane to lithium wire results in a mixture of Ge(Me₃Ge)₄ (1) and (Me₃Ge)₃Ge-Ge(GeMe₃)₃ (3). If alternatively the reaction is completed in −78 °C, expected product tetrakis(trimethylgermyl)germane (1) exclusively obtained (Figure 2.5).

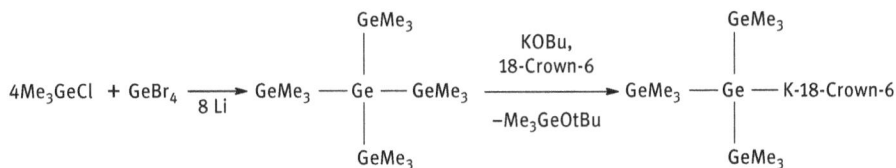

**Figure 2.5:** Synthesis of polygermanes using crown ether.

In analogy with the preparing of KSi(Me₃Si)₃[36 ] and KGe(Me₃Si)₃, the transformation of neopentagermane(1) to the potassium salt of tris(trimethylgermyl)-germanide(2) by potassium tert-butoxide response. The reaction can be efficiently completed in the presence of solvent benzene or toluene along with 1 equivalent of 18-crown-6; on the other hand, it did not proceed so efficiently in THF lacking crown ether. Alternatively, metalation reagent such as potassium diisopropylamide (KDA) was used.

Strong nucleophilic compound (2) readily reacts with the electrophilic compound. The reaction of compound (2) along with 1,1-dibromoethane produced hexakis(trimethylgermyl)digermane (3) [37]. Followed by this, linear pentagermane of bis{tris(trimethylgermyl)germyl}-dimethylgermane (4) is developed from dichlorodimethylgermane. Silylation of (2) was another simple way for the reaction, chlorotriisopropylsilane and chlorotrimethylsilane to acheive o-tetragermanes 5 and 6 (Figure 2.6).

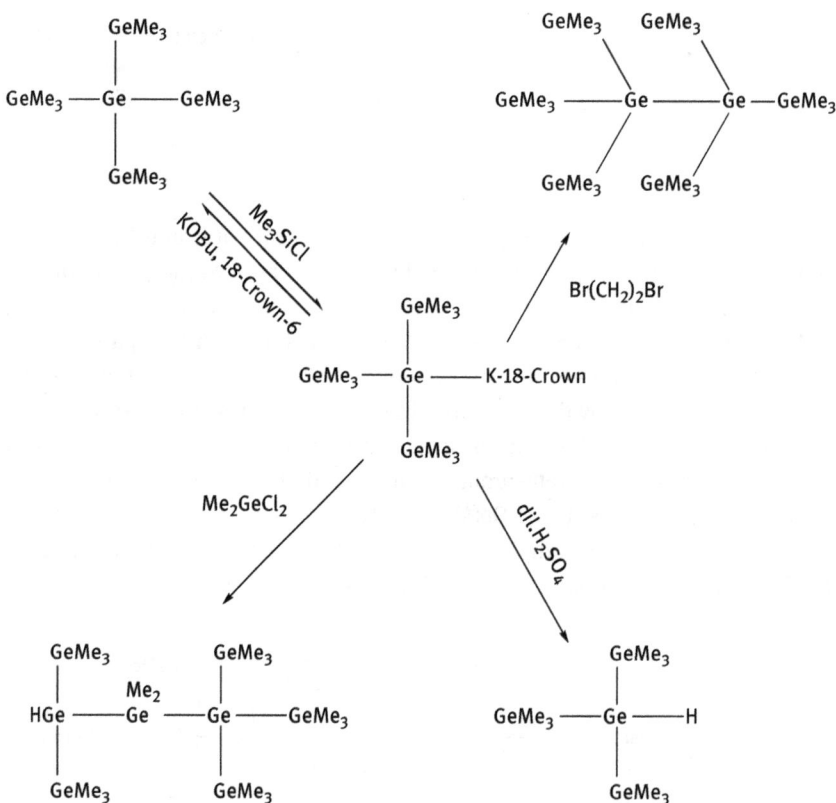

**Figure 2.6:** Reactions of electrophiles along with the crown-ether adduct of $(Me_3Ge)_3GeK$.

The trimethylsilylated compound (5) treated with potassium tert-butoxide further the selectivity of alkoxide attack on trimethylsilyl and trimethylgermyl groups. The fact that an exclusive attack on the trimethylsilyl group was observed is probably due to the stronger Si–O bond in comparison to Ge–O bond (Figure 2.6).

Further, hydrolysis of compound 2 along with dil. $H_2SO_4$ was carried out. Hydrogermane (7) formed a facile way due to the absolute amount of oxygen. Similarly, the reaction between hexakis(trimethylgermyl)digermane (3) and

(Me$_3$Si)$_3$GeK [38] formed mono as well as dianions, which shows a logical further step. However, the reaction of (3) required special solvent conditions (with KO$^t$Bu in either THF or benzene/18-crown-6), especially, the formation of (2) shows anionic species. Mass spectroscopic analysis provided additional evidence for the formation of (Me$_3$Ge)$_3$-GeOtBu. At last, the internal bond breaking of Ge–Ge assumes that high steric accessibility through longer Ge–Ge bonds. The clean formation of K(Me$_3$Si)$_2$Ge–Ge (SiMe$_3$)$_2$K compound by using precursor-like (Me$_3$Si)$_3$Ge–Ge(SiMe$_3$)$_3$ [39]. To obtain 2,2,3,3-tetrakis(trimethylsilyl)hexamethyltetragermane (8) (Scheme 2), 2 eq. of Me$_3$GeCl reacts with K(Me$_3$Si)$_2$GeGe(SiMe$_3$)$_2$K [54]. Compound (8) reacts with compound (2) in the presence of KO$^t$Bu and bond breaking of Si–Ge and then 1, 2-dipotassium compound (9) occurred. Me$_3$GeCl along with germylation used (Me$_3$Si)(Me$_3$Ge)$_2$Ge-Ge (GeMe$_3$)$_2$(SiMe$_3$) (10) as a starting material for the preparation of K(Me$_3$Ge)$_2$GeGe (GeMe$_3$)$_2$K. At the end step of the reaction, di-anionic K(Me$_3$Ge)$_2$Ge-Ge(GeMe$_3$)$_2$K (11) is obtained as crown ether as shown in Figure 2.7.

The preparation of polygermane by electroreduction of Ge–Si and Ge–Ge bond, along with magnesium electrode [40] forms a germane–silane copolymer. Electroreduction of chlorotrimethylgermane (1) formed 84% yield product of hexamethyldigermane (2).

The execution of electroreduction of (1) by the Mg rod as a cathode and on the other hand anode is in an undivided cell under sonication (47 MHz), and an anode and a cathode were 15 s. Moreover, the electroreduction of a mixture of (1) and chlorotriphenylsilane 3a or chlorodimethylphenylsilane 3b (molar ratio of 1:3, 1:2) under similar conditions afforded the corresponding germylsilane 4a or 4b as the product as shown in Figure 2.8.

Electroreductive synthesis method plays a vital role in the formation of Ge–Si and Ge–Ge bonds. The high MW material such as polygamy as well as germane–silane copolymer was designed by the electrochemical reduction. A polygermane is obtained by the electroreduction of dichlorophenylbutylgermane (5) and MW obtained 19,900 for 10% yield. Figure 2.9 shows the formation of germane–silane copolymer by using a mixture of dichlorophenylbutylgermane (5) and dichloromethylphenylsilane (6).

Tetrakis(trimethylgermyl) germane was synthesized by the reaction of tetrabromogermane and chlorotrimethylgermane along with lithium at −78 °C. Reaction with potassium tert-butoxide/18-crown-6 gave (Me$_3$Ge) $_3$GeK 18-crown-6, which showed electrophiles in Figure 2.10. In the reaction of (Me$_3$Ge)$_3$GeSiMe$_3$ with tert-butoxide/18-crown-6, attack at the trimethylsilyl group was reported by Helena et al. [41].

Reactions of 4,4-dichlorodithienogermoles with sodium (Na), by reprecipitation of the organic compounds, afforded poly(dithenogermane-4,4-diyl)s. The absorption edges were at lower energies than that of a monomeric dithienogermole derivative. Polygermanes synthesized by oxidizing and copolymers composed of di-$n$-butylgermane and dithienogermole units revealed that the bathochromic shift (red-shift) absorption is due to conjugation of polygermane backbone σ-orbital with the dithienogermole π-orbital (Figure 2.11) [42].

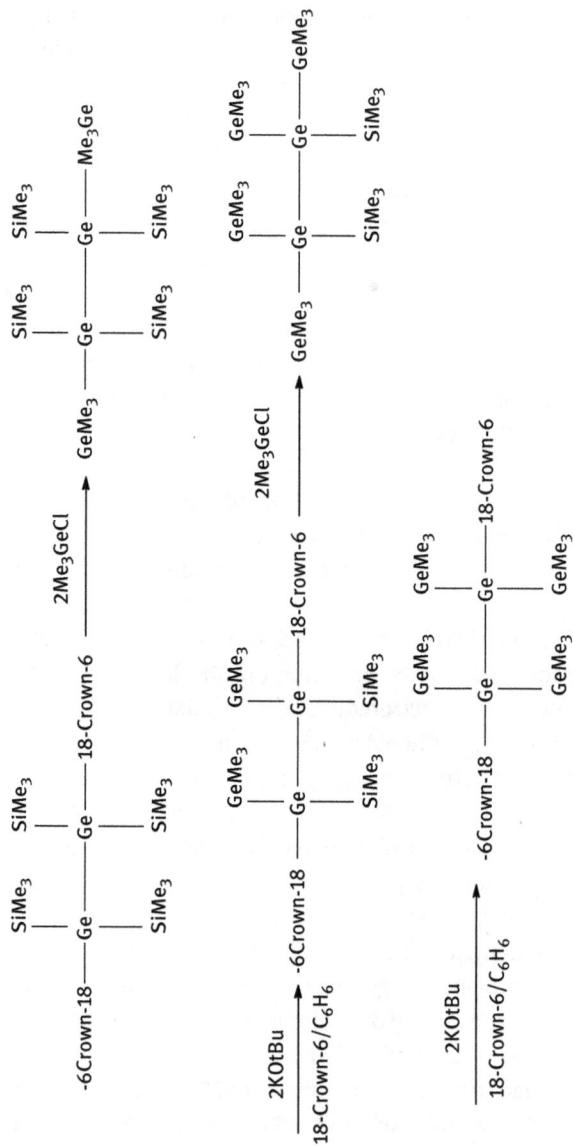

**Figure 2.7:** Successive interchange of trimethylsilyl against trimethylgermyl shows the formation of K(Me₃Ge)₂GeGe (GeMe₃)₂K (11).

$$Me_3GeCl + ClSiPhR_2 \longrightarrow Me_3GeSiPhR_2 + (Me_3Ge)_2$$

a; R = Ph

b; R = Me

**Figure 2.8:** Synthesis of chlorodimethylphenylsilane.

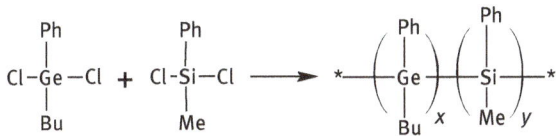

**Figure 2.9:** Synthesis of germane–silane copolymer.

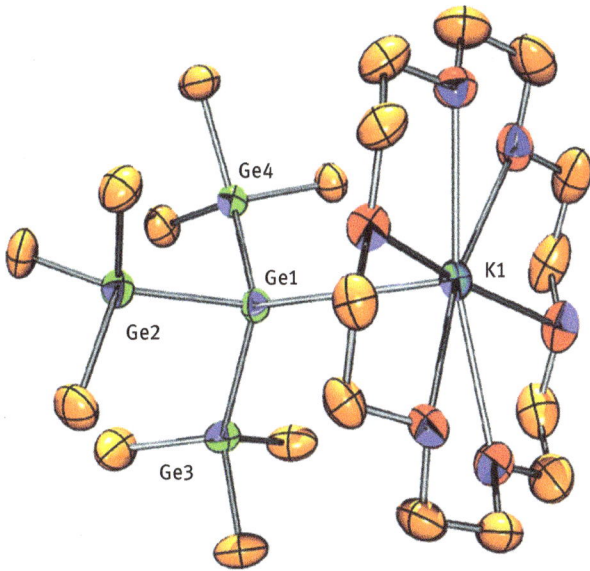

**Figure 2.10:** Building blocks of polygermanes [41].

## 2.3 Properties

Polystannanes are yellow in color and its number average molecular masses and polydispersity index are 10–70 kg/mol and 2–3, respectively. The molar mass of polystannanes can be adjusted by variation of catalyst and reaction temperature. The degree of conversion depends on growth onto the catalyst, for example, by insertion

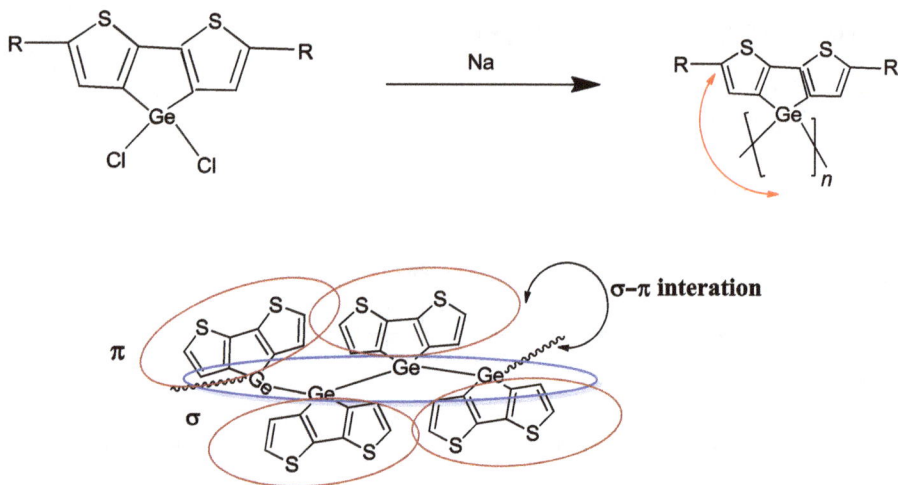

**Figure 2.11:** Synthesis of poly(dithienogermole)s.

of SnR$_2$-like units. Poly(dialkylstannane) is found to be thermotropic and displayed first-order phase transitions from one liquid-crystalline phase into another or directly to the isotropic state, depending on the length of the side groups. More specifically, poly(dibutylstannane) exhibits an endothermic phase transition at freezing temperature from a rectangular to a pure nematic phase. Polystannanes are semiconductive and its temperature-based conductivity is similar to pi-bond-conjugated carbon-based polymers.

Polygermanes, linear polymers of germanium backbone, are the germanium analogues of polyolefin that exhibit properties similar to those of polysilane. Due to their σ-delocalization, they are readily undergoing photoscission and display intense UV absorption maxima, which are 20–40 nm red-shifted from those of the polysilane analogue. Polygermynes appear to adopt a random network structure identical to that of polysilynes. The physical properties of polygermynes are: they are readily soluble in organic solvent, have good film-forming properties, lack of any peak in X-ray diffraction spectra, and lack of any discrete melting point.

## 2.4 Applications

Polystannane is used as a catalyst for organic conversions and also for rubber tube lines due to low glass transition temperature value. Polygermane is employed in solar panel and is useful in optical applications [43–45].

## 2.5 Conclusion

In this chapter, a brief review of polygermanes and polystannanes is discussed in detail. The preparation of polystannanes by various methods, in particular by procedures based on sodium (Na), electrochemical polymerization of dichloro-diorganostannanes; the Wurtz reaction with dichlorodiorganostannanes, and catalytic dehydropolymerization of dihydrodiorganostannanes, have been discussed. Poly(dialkylstannane)s can show liquid-crystalline properties at room temperature. Polygermanes with high MW were synthesized by an advanced Wurtz coupling reaction of dichlorogermanes and sodium (Na) metal, and by a method using Grignard reagents and diiodogermylene having the molecular weight of $10^4$ daltons.

## References

[1] Isaka, H., Fujiki, M., Fujino, M., & Matsumoto, N. A new type of inorganic polymer with ordered SiSiGe sequences, Macromolecules, 1991, 24(9), 2647–2648.

[2] Trefonas, P., & West,, R. Organogermane homopolymers and copolymers with organosilane, J. Polym. Sci., Polym. Chem., 1985, 23(8), 2099–2107.

[3] Shono, T., Kashimura, S., Ishifune, M., & Nishida, R. Electroreductive formation of polysilanes, J. Chem. Soc. Chenz. Commun., 1990, 0(17), 1160–1161.

[4] Samuel, L. M., Sanda, P. N., & Miller, R. D. Thermally stimulated current studies of charge transport in a σ-conjugated polymer, Chem. Phys. Lett., 1989, 159(2–3), 227–230.

[5] Abkowitz, M., Bapler, H., & Stolka, M. Common features in the transport behavior of diverse glassy solids: Exploring the role of disorder, Philos. Mag. B., 1991, 63(1), 201–220.

[6] Yajima, S., Hasegawa, Y., & Hayashi, J. Synthesis of continuous silicon carbide fiber with high tensile strength and high Young's modulus, J. Mater. Sci., 1978, 13(12), 2569–2576.

[7] Wang, Y. D., Feng, C. X., & Song, Y. C. Study on the process of continuous Sic fibers, Aerospace Mater. Tech., 1997, 2, 21–25.

[8] Okamura, K., & Seguchi, T. Application of radiation curing in the preparation of polycarbosilane-derived SiC fibers, J. Inorg. Organomet. Polym., 1992, 2(1), 171–179.

[9] Youngblood, G. E., Jones, R. H., Kohyama, A., & Snead, L. L. Radiation response of SiC-based fibers, J. Nucl. Mater., 1998, 258–263, 1551–1556.

[10] Sharma, H. K., & Pannell, K. H. Tin Chemistry: Fundamentals, Frontiers, and Applications, Eds, Davies, A. G., Gielen, M.,Pannell, K. H.,Tiekink, E. R. T., Wiley, Chichester, 2008, 376–391.

[11] Devylder, N., Hill, M., Molloy, K. C., & Price, G. J. Wurtz synthesis of high molecular weight poly(dibutystannane), Chem. Commun., 1996, 0(6), 711–712.

[12] Imori, T., Lu, V., Cai, H., & Tilley, T. D. Metal-catalyzed dehydropolymerization of secondary stannanes to high-molecular-weight polystannanes, J. Am. Chem. Soc., 1995, 117(40), 9931–9940.

[13] Mustafa, A., Achilleos, M., Ruiz-Iban, J., Davies, J., Benfield, R. E., Jones, R. G., Grandjean, D., & Holder, S. J. Synthesis and structural characterisation of various organosilane–Organogermane and organosilane–Organostannane statistical copolymers by the Wurtz reductive coupling polymerisation: 119Sn NMR and EXAFS characterisation of the stannane copolymers, React. Funct. Polym., 2006, 66(1), 123–135.

[14] Miles, D., Burrow, T., Lough, A., & Foucher, D. Wurtz coupling of perfluorinated dichlorostannanes, J. Inorg. Organomet. Polym., 2010, 20(3), 544–553.

[15] Jones, R. G., & Holder, S. J. Review high-yield controlled syntheses of polysilanes by the Wurtz-type reductive coupling reaction, Polym. Int., 2006, 55(7), 711–718.

[16] Bratton, D., Holder, S. J., Jones, R.G., & Wong, W. K. C. The role of oligomers in the synthesis of polysilanes by the Wurtz reductive coupling reaction, J. Organomet. Chem., 2003, 685(1–2), 60–64.

[17] Trummer, M., & Caseri, W. Diorganostannidedianions ($R_2Sn_{2}-$) as reaction intermediates revisited: In situ $^{119}$Sn NMR studies in liquid ammonia, Organometallics, 2010, 29(17), 3862–3867.

[18] Trummer, M., Zempp, J., Sax, C., & Caseri, W. Reaction products of dichlorodiorganostannanes with sodium in liquid ammonia: In-situ investigations with ($^{119}$)Sn NMR spectroscopy and usage as intermediates for the synthesis of tetraorganostannanes, J. Organomet. Chem., 2011, 696(19), 3041–3049.

[19] Trummer, M., Choffat, F., Rämi, M., Smith, P., & Caseri, W. Polystannanes – synthesis and properties, Phosphorous Sulfur Silicon. Relat. Elem., 2011, 186(6), 1330–1332.

[20] Okano, M., Matsumoto, N., Arakawa, M., Tsuruta, T., & Hamano, H. Electrochemical synthesis of dialkyl substituted polystannanes and their properties, Chem. Commun., 1998, 0(17), 1799–1800.

[21] Okano, M., & Watanabe, K. Electrochemical synthesis of stannane–Silane, and stannane–Germane copolymers, Electrochem. Commun., 2000, 2(7), 471–474.

[22] Okano, M., Watanabe, K., & Totsuka, S. Electrochemical synthesis of network polystannanes, Electrochemistry, 2003, 71(4), 257–259.

[23] Thompson, S. M., & Schubert, U. Dehydrogenerative stannane coupling by platinum complexes, Inorg. Chim. Acta., 2003, 350, 329–338.

[24] Woo, H. G., Park, J. M., Song, S. J., Yang, S. Y., Kim, I. S., & Kim, W. G. Catalytic Dehydropolymerization of Di-n-butylstannane n-Bu2SnH2 by Group 4 and 6 Transition Metal Complexes, Bull. Korean Chem. Soc., 1997, 18(12), 1291–1295.

[25] Woo, H. G., Song, S. J., & Kim, B. H. Redistribution/dehydrocoupling of tertiary alkylstannane in n-Bu$_3$SnH catalyzed by group 4 and 6 transition metal complexes, Bull. Korean Chem. Soc., 1998, 19(11), 1161–1164.

[26] Choffat, F., Schmid, D., Caseri, W., Wolfer, P., & Smith, P. Synthesis and characterization of linear poly(dialkylstannane)s, Macromolecules, 2007, 40(22), 7878–7889.

[27] Choffat, F., Smith, P., & Caseri, W. Polystannanes: Polymers of a molecular, jacketed metal–Wire structure, Adv. Mater., 2008, 20(11), 2225–2229.

[28] Babcock, J. R., & Sita, L. R. Highly Branched, High molecular weight polystannane from dibutylstannane via a novel dehydropolymerization/rearrangement process, J. Am. Chem. Soc., 1996, 118(49), 12481–12482.

[29] Brook, A. G., Abdesaken, F., & Scollradl, H. Synthesis of some tris(trimethylsilyl)germyl compounds, J. Organomet. Chem., 1986, 299(1), 9–13.

[30] Fischer, J., Baumgartner, J., & Marschner, C. Silylgermylpotassium compounds, Organometallics, 2005, 24(6), 1263–1268.

[31] Baines, K., Mueller, K. A., & Sham, T. K. Tetrakis(trimethylgermyl)silane and Tris (trimethylgermyl)silyllithium, Can. J. Chem., 1992, 70(12), 2884–2886.

[32] Barrau, J., Rima, G., & El, Amine. Satge, Synthesis of Chlorotrimethylgermane, chloropentamethyldigermane and 1,2-dichlorotetramethyldigermane, J. Synth. React. Inorg. Met.-Org. Chem., 1998, 18(1), 21–28.

[33] Herrmann, W. A., Brauer, G. Eds Synthetic Methods of Organometallic and Inorganic Chemistry, Thieme: Stuttgart, 1996, Vol. 2, 251.

[34] Brauer, G. Modified Procedure of Handbuch der preaparative nanorganischen chemie, Ed, F. EnkeVerlag, Stuttgart, 1978, 723–726.

[35]   Glockling, F., Light, J. R. C., & Strafford, R. G. Alkylpolygermanes, J. Chem. Soc. (A), 1970,
       426–432.

[36]   Gilman, H., Holmes, J. M., & Smith, C. L. Branched-chain methylated polysilanes containing a
       silyl-lithium group, Chem. Ind. (London), 1965, 848–849.

[37]   Gilman, H., & Smith,, C. L. Tris(trimethylsilyl)silyllithium, J. Organomet. Chem., 1968, 14,
       91–101.

[38]   Marschner, C. A new and easy route to polysilanylpotassium compounds, Eur. J. Inorg.
       Chem., 1998, 1998(2), 221–226.

[39]   Gilman, H., & Harrell, R. L. For the analogous synthesis of hexakis(trimethylsilyl)disilane, J.
       Organomet. Chem., 1967, 9(1), 67–76.

[40]   Castel, A., Riviere, P., Saint-Roch, B., Stage, J., & Malrieu, J. P. Synthese et etude spectraleuv
       de chaines polymetallees (Ge, Si); analyse theorique des effects de substitution, J.
       Organornet. Chem., 1983, 247(2), 149–160.

[41]   Hlina, J., Baumgartner, J., & Marschner, C. Polygermane building blocks, Organometallics,
       2010, 29(21), 5289–5295.

[42]   Nakamura, M., Ooyama, Y., Hayakawa, S., Nishino, M., & Ohshita, J. Synthesis of poly
       (dithienogermole)s, Organometallics, 2016, 35(14), 2333–2338.

[43]   Yukimoto, Y., & Aiga, M. A SiGe: Alloy and its application to tandem type solar cell, MRs
       Proceedings, 1986, 70, 493.

[44]   Rivard, E. Inorganic and organometallic polymers, Annual Reports Section" A"(Inorganic
       Chemistry), 2010, 106, 391–409.

[45]   Pfister, G., & Scher, H. Dispersive (non-Gaussian) transient transport in disordered solids,
       Adv. Phys., 1978, 27(5), 747–798.

# 3 Phosphorus-based polymers: polyphosphate, polyphosphoric acids, phosphonate, and polyphosphazene

**Abstract:** Polyphosphate (poly P) is one of the several molecules on the Earth that effectively store energy within their covalent bonds, Polyphosphate (polyp) is an ester of many structural units of tetrahedral phosphate connected together by high-energy phosphoanhydride bonds. It is found in high abundance in living and non-living things. In this chapter, we discuss about the chemistry of polyphosphates and their applications for the material community in various purposes like water softening agent, composite materials in engineering, flame retardant, antidrug delivery for cancer, high-performance polymer, biomedical field, food industry, and amphiphilic grafted hyperbranched polymers.

**Keywords:** inorganic, polyphosphate, energy, functionalization, industrial, drug delivery, composite

## 3.1 Introduction

Inorganic polyphosphate (poly P) are abundantly found in living organisms. They are linear molecules in an ionic form, which are bonded via a high-energy phosphoanhydride bond [1]. Phosphate with a single unit molecule is called simple phosphate or orthophosphate. Phosphate with two, three, or more units is called as pyrophosphates, tripolyphosphates, and polyphosphates. It may be linear or cyclic in structure as shown in Figure 3.1.

Polyphosphates are the source of energy for biotic and probiotic like Lactobacillus. These microbes confer the benefit to the health of host when present in adequate amounts [2]. Unfortunately, in today time no obits polyphosphate can be found on Earth. It is present in marine sediments underlying anoxic versus oxic bottom waters. Further, it gets mixed due to bioturbation and ventilated by burrowing, bioirrigation [3]. Biogenic phosphorus from the wetland, pond sediment [4], and waterbodies [5] have phosphate in the form of polyphosphate, orthophosphate, and pyrophosphate. Various environmental compounds such as adenosine 5′-triphosphate sodium salt, guanosine 5′-triphosphate sodium salt, β-nicotinamide adenine dinucleotide, dicalcium pyrophosphate, tripolyphosphatepentasodium salt anhydrous [6], soil, arable soils, and grassland [7] have phosphate and its derivatives.

Polyphosphate-based compounds such as ammonium polyphosphate (APP), aluminum diethylphosphinate (ADP), octaphenyl polyhedral oligomeric silsesquioxane (OPS) [8], and melamine polyphosphate [9] have properties like flame retardant (FR) [10], tensile strength, modulus of elasticity, limiting oxygen, and water absorbance

https://doi.org/10.1515/9781501514609-004

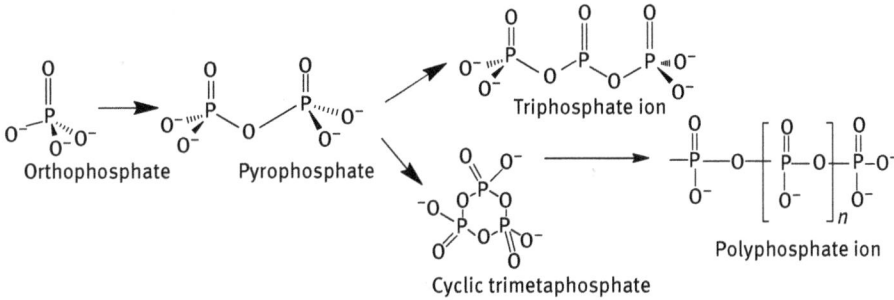

**Figure 3.1:** Structure from orthophosphate to polyphosphate.

[11] to polypropylene [12], polylactic acid [13], and acrylic materials [14] (Table 3.1). It is a chemical water treatment used to remove inorganic contaminant in groundwater, and to maintain or stabilize water quality by reducing corrosion, biofilm, and so on. Polyphosphate is an important component for the growth and production of agriculture and has application in food industries also. It has been used in the medical field in prefabricated bone graft materials, reducing dental hypersensitivity, prevent erosion of teeth, decrease dissolution of hydroxyapatite, pulmonary and nasal transmucosal delivery of macromolecules, oral vaccine, drug delivery, and others.

In this chapter, we have discussed structure-based applications of polyphosphate, which promote the development of interdisciplinary research between the polymer chemistry and material science.

Naturally, phosphate is found as poly(alkylene phosphates) in living organisms to form nucleic acids. The phosphoric acid reacts with polyoils or diepoxides to form poly(alkylene phosphates). Thus, phosphoric acid is a precursor of polymeric phosphates synthesized by either chemical or enzymatic routes [15]. Polyp such as sodium triphosphate (STP, $Na_5P_3O_{10}$) and sodium hexametaphosphate (SHMP, $Na_{15}P_{13}O_{40}$–$Na_{20}P_{18}O_{40}$) influence the proliferation, odontoblastic differentiation, and angiogenic potential of human dental pulp cells [16].

### 3.1.1 Polyp in prokaryotes organism

Polyp as planning has a major role in the evolution of cell on earth. Since prebiotic evolution, polyphosphate has been considered as an energy reservoir and most probable compound for RNA, DNA, and proteins [17]. Therefore, it is abundantly found in the cell in nature and found to have a crucial role in the origin and survival of organic species [18]. Furthermore, it also has the capability of cell growth and their responses toward stresses and stringencies, and the virulence of pathogens. In prokaryotic organisms such as *Escherichia coli*, an enzyme like polyphosphate kinase (PPK) is

**Table 3.1:** Different polyphosphates and their uses.

| Name of polyphosphate | Chemical formula | Structure | Properties and applications |
|---|---|---|---|
| Polyphosphoric acid | $H_3O_4P$ | | Used in organic synthesis for cyclizations and acylations |
| Sodium trimetaphosphate | $Na_3P_3O_9$ | | Used in food and construction industries |
| Sodium hexametaphosphate | $Na_6P_6O_{18}$ | | Used as a sequestrant and as a food additive |

| | | | |
|---|---|---|---|
| Ammonium polyphosphate | $[NH_4PO_3]_n$ | (structure) | Used as a food additive, emulsifier, fertilizer, and flame retardant |
| Aluminum diethylphosphinate | $((C_2H_5)_2PO_2)_3Al$ | (structure) | Used as a flame retardant in engineering plastics |
| Melamine polyphosphate | $(C_3H_9N_6O_4P)_n$ | (structure) | Environmental protection-type nonhalogen flame retardant |

present: polyphosphate is prepared from ATP (adinosine triphosphate) and E. coli exopolyphosphate.

Polymer length depends on various organisms and their physiological stage. Yeast exopolyphosphatase produced in excess of *E. coli* leads to distortion of PPK gene as it increases sensitivity toward $H_2O_2$ and heat shock that work as the mutant source. Hence, it reduces rpoS expression, which is a regulatory network that governs the expression of stationary-phase-induced genes. Besides this polyp also governs stress-inducible genes, which is not directly regulated by repose expression [21]. Kim et al. [22] cloned a pathogenic gene ppk in *Shigella flexneri*, *Salmonella enterica* serovar Dublin, and *Salmonella enterica* serovar Typhimurium and observed the decrease in growth, responses to stress and starvation, loss of viability, polymyxin sensitivity, sensitivity toward acid and heat, and loss of invasiveness in epithelial cells. Furthermore, Rao et al. [18] classified PPK into PPK1 and PPK2 and reported that mutant enzyme leads to defect in motility, quorum sensing, biofilm formation, and virulence. The ATP site of the PPK1 gets adversely affected even in the presence of a minute amount of inhibitor. The PPK2 has distinctive kinetic properties and also affects the virulence of species.

Kornberg et al. [23] reviewed the function of conserved enzyme PPk (also referred to as ADP phosphotransferase), which is encoded by ppk gene Pk, which converts polyp and ADP into ATP in many bacterial or pathogenic species such as *Vibrio cholerae, Pseudomonas aeruginosa, Mycobacterium tuberculosis, Yersinia pestis, Helicobacter pylori, Salmonella typhimurium, Shigella flexneri, Bordetella pertussis, Neisseria meningitides*, and *Escherichia coli*. Whereas exo- and endo-polyphosphatases have been identified and isolated as synthetic unit of poly P or its physiologic functions unit in yeast and mammal cells.

Thus, these enzymes may form or degrade polyp. On the basis of its quantitative availability of polyp, its function varied and depends on chain length, biologic source, and subcellular location as follows [24]:
- energy supply and ATP substitute,
- a reservoir for Pi,
- a chelator of metals,
- a buffer against alkali,
- a channel for DNA entry,
- a cell capsule, and
- a regulator of responses to stresses and adjustments for survival in the stationary phase of cultural growth and development.

### 3.1.2 Polyp in eukaryotes organism

Polyp has different roles in the pathological and physiological function of higher eukaryotes. It has less function in mammalian cells as compared to microorganisms [25].

The activated platelet secretes more than 300 active substances that are very important for hemostasis and coagulation. Polyphosphate is a highly anionic linear polymer that is produced by ATP in enzymatic reaction and released by activating platelets [26] or linked by ATP bonds, which put their influence on blood plasma by binding to several proteins such as factor XII, fibrin(ogen), trombone, and factor VII-activating protease. Hence, some proteins are polyp binding factors [27–29]. PolyP is considered as a potent prothrombotic and proinflammatory agent [30]. Calcium-containing poly (with approximately 100 monomers) accelerates the activation of proenzyme thrombin-activatable fibrinolysis inhibitor (TAFI) by thrombin and plasmin. Whereas polyp with sodium counterions such as Na-PolyP700, Na-PolyP100, and Na-PolyP70 enhance only by plasmin for the activation of TAFI by Plug and Meijers [31].

*Phragmatopoma californica* (Fewkes) creates glue in seawater, which consist of polyphosphates, polysulfates, and polyamines. Polyphosphate subgranules of the heterogeneous adhesive packets help in covalently cross-links adhesion of sand grains and biomineral particles to form well-defined tubular shells reported by Stewart et al. [32]. *Euglena gracilis* has high concentration polyphosphate as intracellular metal ion chelators [33].

### 3.1.3 PolyP from abiotic component and chemical pathway

Phosphorus mines produce waste rock when dumped in the river, then on the basis of pH their leaching takes place. At lower pH, leaching takes place at a higher rate than the neutral condition. Also, when environmental pH turns from weak alkalinity to slight acidity, release of phosphorus gets increased. In the leaching process, total phosphorus consists of orthophosphates, polyphosphates, and organic phosphates. Hence, leaching phosphorus mining from waste rock releases polyphosphorus [34].

Inorganic polyphosphate salts that have a high-temperature, cost-efficient lubricant can be prepared by a hot metal process, providing desired friction and wear characteristics [35]. Zinc dialkyl dithiophosphate (ZDDP) and blend have been used as the main antiwear lubricant additive, including dispersants, antiwear additives, antioxidants, defoamants, corrosion inhibitors, viscosity modifiers, and pour point depressants. ZDDP, ashless fluorothiophosphates and thiophosphates blend lead to the formation of tribofilm with only short-chain polyphosphates of zinc [36]. But the use of ionic liquids in base oil gives longer chain polyphosphate. Length change occurs due to higher levels of networking. Both ZDDP and ionic liquid have a different role when used as primary cationic species present in the polyphosphate network [37]. These roles are:

- When ZDDP is used, it replaces Zn and Fe
- Whereas Ca replaces Fe, when ionic liquids are used

Phosphorus-containing polymers, and particularly poly(phosphoester)s (PPEs), are promising materials for biomedical applications [38, 39]. The pentavalent phosphorus particle permits the plan of secluded structures, and the inborn ester bonds in the polymer spine make them hydrolytically degradable. Moreover, water-dissolvable PPEs are the promising possibilities of medication conveyance vehicles [40–42] due to their "stealth impact," like poly(ethylene glycol) (PEG), while their degradability keeps any potential bioaccumulation [43, 44]. Through exact control of their science, the physical properties, debasement items, and time can be tuned [45–47].

Poly(phosphonate)s (PPE) contain a synthetically steady P–C security, supplanting one of the P–O–C obligations of poly(phosphate)s, which impacts hydrolysis rates, as the P–C security is steady against hydrolysis; however, microorganisms can divide the phosphonate linkage [48].

Be that as it may, poly(phosphonate)s are predominantly found as fragrant oligomers arranged by step-development polymerization and not very many water solvent, all-around characterized precedents have been accounted for [49–51]. In poly(phosphonate)s the P–C bond is ordinarily introduced as a pendant gathering; in any case, uncommon chain P–C linkages have been reported [47, 52, 53]. The primary ring-opening polymerization (ROP) of 2-alkoxy-2-oxo-1,3-oxaphospholanes with both ethyl and butyl side chains has conveyed the P–C bond inside the ring structure and in this manner shaping in-chain poly(phosphonate)s upon polymerization. These supposed phostones have been of some enthusiasm in the past because of their potential applications as glycomimetics [54–56].

Poly(organo)phosphazenes belong to the family of inorganic-based hybrid polymers with very diverse properties because of the possibilities of many organic substitutes. Its backbone consists of alternating nitrogen and phosphorus atoms, where organic substituents are linked to phosphorus atoms as side groups. Polyphosphazenes themselves have a long history with cross-linked elastomeric materials ("inorganic rubber") consisting of phosphorus and nitrogen. The isolation of soluble poly(dichloro)phosphazene [NPCl$_2$] was first reported by Allcock and coworkers [57].

The bonding in phosphazenes is quite different as compared to the other organic polymers such as polyethylene and polyisoprene. In phosphazene, both phosphorus and nitrogen atoms contribute five valence electrons per repeating unit and if two of the electrons from nitrogen are considered to locate at lone-pair orbital and electron pairs are assigned to the sigma bond for polymer framework but still two electrons are left, one from phosphorus and one from nitrogen. Both single electrons do not remain unpaired. It is supposed that the electron present on nitrogen is accommodated in a 2p$_z$ orbital, whereas one unpaired electron of phosphorus in a 3d orbital to form a different arrangement (Figure 3.2). Thus, although the pi-bonds are delocalized over "islands" of three skeletal atoms, they are not broadly delocalized over the whole chain because of the orbital mismatch and the nodes that occur at every phosphorus. Therefore, phosphorus atoms present in polyphosphazene can use as

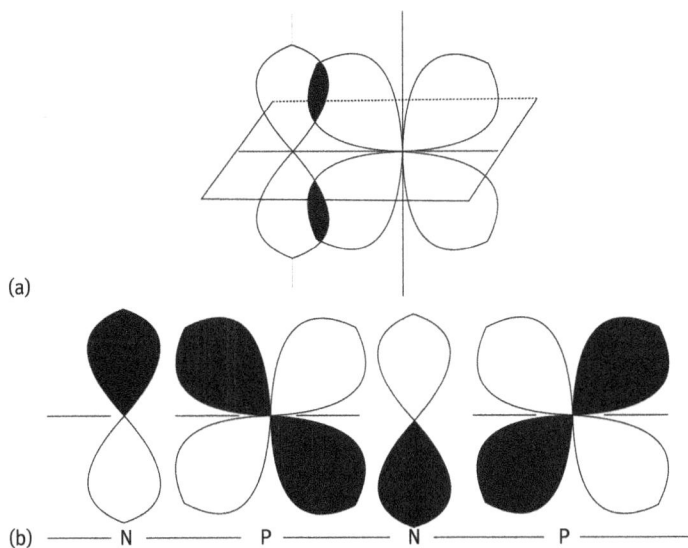

(a)

(b) —— N —————— P —————— N —————— P ——————

**Figure 3.2:** (a) Typical d$\pi$ – p$\pi$ bonding and (b) delocalization of $\pi$ electron cloud on -N-P-N-P-bond.

maximum as possible d orbitals ($d_{xy}$, $d_{yz}$, $d_{zx}$, $d_{x2-y2}$, and $d_{z2}$). As a result, torsion of phosphorus–nitrogen bond can take place, which brings the proper position of p orbitals of nitrogen with the d orbitals of phosphorus with respect to any torsion angle. Hence, the inherent torsional barrier is much smaller than in a p$\pi$–p$\pi$ double bond of the type found in organic molecules. Advanced research about this polymer reveals that the inherent torsional barrier in the principal chain was found to be lesser than 0.1 kcal per bond [58, 59].

The use of 3d$\pi$–2p$\pi$ bonding model is a very important indication to explain the physical–chemical properties of phosphazene [60–62]. The effective 3d$\pi$–2p$\pi$ bonding, phosphorus 3d orbitals ($d_{xz}$ and $d_{yz}$), and nitrogen $2p_z$ orbital have appropriate symmetry to form very effective overlapping. If only $d_{xz}$ orbital involves forming 3d$\pi$–2p$\pi$ bonding, then a broad, delocalized heteromorphic pseudoaromatic $\pi$ orbital could be formed but if $d_{yz}$ is involved then by using a symmetry approach a homomorphic $\pi$ system is produced. (It is comparable to the p$\pi$–p$\pi$ aromatic system.)

If orbitals participate equally, then $\pi$ orbital should be separated into an island of $\pi$ character, interrupted at each phosphorus atom and broad delocalization effect is not observed. ESR study suggests that there is no torsional barrier at P–N bond, and the composition of $\pi$ cloud of phosphorus includes 3d with a substantial amount of second and fourth shell orbital mixing. Π bonding between phosphorus and ligand group is also possible in phosphorus and it will employ $d_z{}^2$. The other possibility is that the lone pair donates to phosphorus to form a coordinate $\pi'$ bond. The remaining $d_{x^2-y^2}$ and $d_{xy}$ are available to accept this lone pair.

The conformation assumed by a polyphosphazene can be understood mainly in terms of the repulsions or attractions between side groups attached to the nearby phosphorus atoms. It is shown in Figure 3.3 that the side group attached to phosphorus atoms tries to move away as maximum as possible in this transplanar conformation. Hence, this confirmation should minimize internal repulsions and generate the lowest energy. Molecular mechanic calculations tend to confirm this supposition.

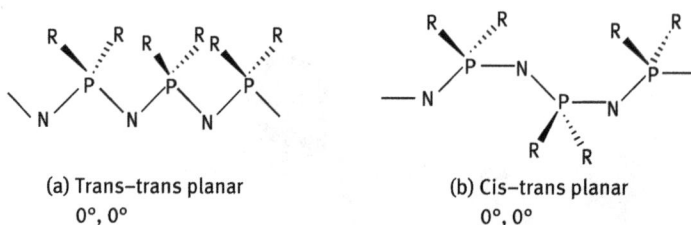

(a) Trans–trans planar
0°, 0°

(b) Cis–trans planar
0°, 0°

**Figure 3.3:** Stereochemistry of polyphosphazene structure: (a) trans–trans and (b) cis–trans.

Phosphoric acid is a weak acid having a general formula $H_3PO_4$. Phosphoric acid is also known as orthophosphoric acid or phosphoric (V) acid. The prefix ortho- is used to distinguish the acid from related phosphoric acids called polyphosphoric acids (PPAs). Orthophosphoric acid is a nontoxic acid. In its pure form, it is present in solid form. A salt of phosphoric acid is known as phosphates.

## 3.2 Synthetic methods

### 3.2.1 Polyphosphate (polyp)

Sun et al. [63] prepared dinucleoside tri-, tetra-, and pentaphosphates, and their phosphonate by using 4, 5-dicyanoimidazole-promoted tandem P–O coupling reactions in one pot.

In the last two decades, rare earth-doped phosphates have been used for information capacity, white-light-emitting diodes [64], lighting of minilasers, quantum electronics, fiberoptics [65, 66], laser technology [67], and dermatology and dentistry (Er:YAG laser) [68]. Concentrate quenching is also one of the factors for the fabrication of rare earth metal-based compounds. Therefore, various rare earth polyphosphates have been synthesized because of their technological and commercial applications [69–71]. The addition of lanthanide ion in polyphosphate such as $LiEu_xLa_{1-x}(PO_3)_4$ [72] helps in the acceleration of concentration of luminescent ions without significant emission quenching. The polyp has been activated by different alkali lanthanide phosphates with a general formula as $MI.Ln^{III}(PO_3)_4$, where MI represents alkali metal ions and $Ln^{III}$ represents lanthanide ions, which have been investigated due to features like stability under normal conditions of temperature and humidity. Polyphosphate of $KLa_{(1-x)}Dy_x(PO_3)_4$ [73], $KEr(PO_3)_4$ [74], $Lila_{(1-x)}Yb_x$ $(PO_3)_4$ [75], $Eu^{3+}$-doped LiCe $(PO_3)_4$ [76] polyphosphates were synthesized by solid-state technique called flux method.

Ben Hassen et al. [77] synthesized $Cm(PO_3)_4$ by flux method. The reaction between samarium oxide ($Sm_2O_3$), cesium carbonates ($Cs_2CO_3$), and phosphoric acid ($H_3PO_4$, 85%) was carried out at 573 K in a platinum crucible and kept the solution at the same temperature for around a week (equation is given below). Finally, the compound was separated by boiling water.

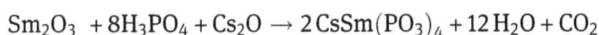

$$Sm_2O_3 + 8H_3PO_4 + Cs_2O \rightarrow 2\,CsSm(PO_3)_4 + 12\,H_2O + CO_2$$

As-prepared polyphosphate has an infinite three-dimensional structure consisting of the double spiral ($PO_3$) in chains linked with neighboring $SmO_8$ and $CsO_{11}$ polyhedral.

Qin et al. [78] modified the surface of APP by using vinyltrimethoxysilane to make it water resistant, increasing dispersities, flame retardancy, and compatibility with the flammable polymer matrix. It has been observed that P–O–N gets partially substituted by P–O–Si. Because of the same P–O–Si bond, it degrades at a higher temperature, thereby increasing thermal stability. Furthermore, polypropylene and dipentaerythritol composite was blended with modified APP by melt blending and extrusion in a twin-screw extruder. The study of char residue showed the formation of complete and compact char, which protects polymer from further combustion while burning.

Szymusiak et al. [79] utilized the conjugation scheme for functionalization of platelet-sized polyp by coating them on the surface of gold nanoparticles. It involves the phosphoramidation of the terminal phosphate of polyp to cystamine in the presence of catalyst EDAC (N-(3-(dimethylamino)propyl)-N'-ethylcarbodiimide hydrochloride). The decrease in disulfide moiety and enhanced covalent anchoring of human pooled normal plasma (PNP) leads to robust contact on a colloidal surface to cease bleeding. It also has the same mobilize factor XII and its coactivating proteins as that in very-long-chain polyp-containing bacteria.

The use of EDAC allows the covalent attachment of primary amines to the terminal phosphate of polyp through stable phosphoramidate linkages. The linkage retains its ability to trigger blood clotting and protects it from exo polyphosphatase degradation, which makes it an excellent probe for biological roles [80].

Reactive oxidants are synthesized by zero-valent iron (nZVI) and ferrous ion (Fe(II)) nanoparticulate in the presence of oxygen. The reaction is accelerated by tetrapolyphosphate as the chelating agent and showed improved yield. nZVI together with TPP has potential application treating groundwater due to the generation of hydroxyl radical. Thus, the function of the polyphosphate is to enhance oxidant production [81]. Iwasaki and Akiyoshi [82] prepared polyp by using 2-iso-propyl-2-oxo-1,3,2-dioxaphospholane, 2-(2-oxo-1,3,2-dioxaphosphoroyloxyethyl-2-bromoisobutyrate), and 2-choresteryl-2-oxo-1,3,2-dioxaphospholane using triisobutylaluminum as an initiator via an ROP. It was further grafted with cholesterol esters 2-methacryloyloxyethyl phosphorylcholine through atom transfer radical polymerization in ethanol to synthesize amphiphilic polyphosphate graft copolymers. The cholesterol groups accelerate the solubility of paclitaxel in an aqueous solution and show cytotoxicity properties.

The hydrophobicity, thermal stability, and shell mechanical properties of polyp are modified by coating AAP with boron containing phenolic resin through in situ polymerization. Further, it was treated with nano-sized silica gel via a thermal cross-linking. As-prepared sample has APP microcapsule with a Si–B containing phenolic resin shell [83].

The self-condensation ROP is carried out in preparation of polyp by using a 2-(2-hydroxyethoxy)ethoxy-2-oxo-1,3,2-dioxaphospholane. The hyperbranched, hydrophilic, biocompatibility, and the biodegradable polyp depicted in Figure 3.4. was used as a drug carrier in vitro condition [84]. The aqueous solution of calcium and magnesium helps in controlling the size of the polyp by precipitation method. It is a reversible process form polyp spherical shape nanoparticles of the diameter of 200–250 nm. It has a uniform electron density, closely resembling the content of acidocalcisomes. Such nanoparticles have procoagulant effects [85].

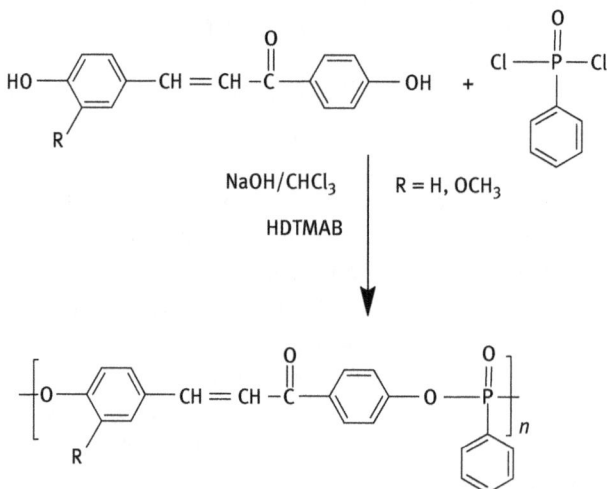

**Figure 3.4:** Synthesis of homopolymers.

Consider 20 mL of NaOH (1 N) arrangement containing a significant amount of hexa decyltrimethyl ammonium bromide (HDTMAB). To this blend, 25 mL of chloroform

containing phenylphophonic dichloride (0.01mol) was added in drops with viva-
cious mixing at 20 °C. After 30 min of blending, the polymer formed between the
natural and fluid layers was isolated and dried in vacuum.

Copolymerization: The copolymers are set up by utilizing the accompanying
procedure. Appropriate diol monomer (0.02 mL) breaks down into 40 mL of 1 N
NaOH solution. A squeeze of HDTMAB is added and mixed well to shape froth.
Chloroform arrangements containing phenyl phosphonic dichloride (0.01 mol) and
terephthaloyl chloride (0.01 mol) researchers prepared independently and at the
same time added to the soluble arrangement of dual monomers utilizing option
pipes at 20 °C with steady stirring. Stirring is continued for up to 30 min and then
the polymer accelerated is filtered, washed with water and methanol, and dried to a
consistent weight in a vacuum at 50 °C. Polymers with various comonomer proportions are likewise arranged along these lines (Figure 3.5).

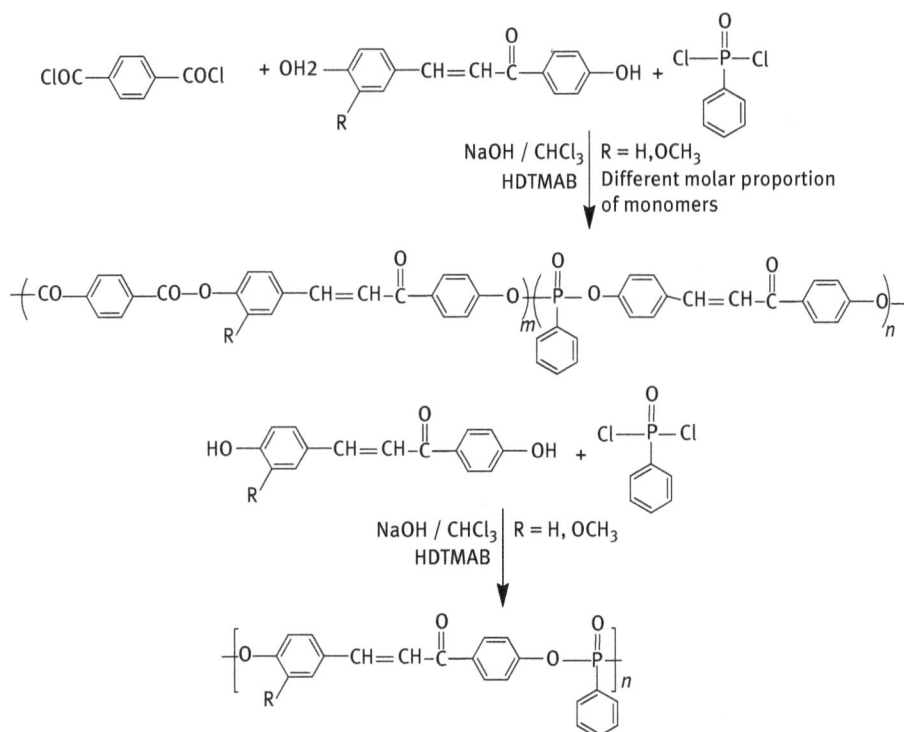

**Figure 3.5:** Synthesis of copolymers.

New polymers and phosphonates have connection with nanomedicine [86], thanks to
their high substance soundness and low toxicity. The basic connection between regular
mixes and their ability to chelate different metal ions represents the base of the ene-
mies of metabolite movement and enables them to go after active sites of compounds

or cellular receptors [87]. Monteil and associates connected a system to orchestrate bisphosphonates having functioned PEG side chains with a particular length with the end goal to structure a novel class of half breed nanomaterials created by tetraphosphonate-complex – gold COOH-ended PEG-coated NPs (bis-POPEG–AuNPs) [88].

In material science, the most versatile class material is known as PPEs, because of their measured union and expansive scope in various conceivable uses [89]. The explicit PPE is DNA, the stockpiling of organic data, the reason for life. In research laboratory, strong stage oligonucleotide blend or polymer as a chain response in anomaly is utilized for the amalgamation of DNA segments [90]. A few highlights provide this phosphorus (P)-based polyesters to exceptionally encouraging material science research world Initially in real contrast to flexible polyesters destructed by carboxylic, that structure modified by using pentavalent phosphorus. This permits functional acids labeling and then degradability of polyesters studied by Penczek and coworkers. Further, polyphosphites and polyphosphates are developed by researchers, using ROP and polycondensation method.

An exceptionally fascinating subclass is poly(phosphonate)s after overlooked them in the scholarly world in spite of some very encouraging fire impeding properties of badly characterized oligomeric poly(phosphonate)s. In poly(phosphonate)s, two molecules of phosphoesters develop them in a chain polyester, while the side chain depends on specifically connected alkyl or aryl functional groups (polyesters of alkyl- or aryl-phosphonic acids, this makes them normally more steady than the relating polyphosphates with three ester groups [91].

Novel procedures to combine PPEs in a controlled way have been researched as of late. Metathesis is acyclic diene methathesis (ADMET) polymerization and ring-opening metathesis polymerization, respectively [92–94]. However, the anionic ring opening of cyclic phosphates, so-called phospholanes, has discovered particles of ultra-quick natural synergist frameworks that have been created by Iwasaki and are investigated in some ongoing research works by Wooley and coworkers [95–98]. In spite of the high control toward the start of the polymerization, conversion is restricted to about half as transesterification responses turned out to be overwhelming at higher transformations and expand the atomic weight appropriation distinctively. This results in untreated and for the most part nonrecoverable monomer loss. Only for sterically demanding [95] or practical alkoxy residues [96], conversions as high as 80–90% are reported, yielding polymers with low subatomic dispersities. Similarly, very thin subatomic weight conveyances for polyphosphoramidates are obtained for changes under 68% [98].

The primary living ROP of a cyclic phosphate, that is, 2-methyl-1,3,2-dioxaphospholane2-oxide (Me EP, 2) is a powerful convention of the union of exceedingly water-dissolvable PPEs. For the principal time, high atomic weight poly(phosphonate)s are open with restricted subatomic weight conveyances by means of a chain development mechanism. Up to high conversions (above 90%), no articulated transesterification is observed, yielding hydrolytically degradable polymers with promising applications in the biomedical field and materials science. Mauldin and collaborators have combined a polyphosphonate by liquefying buildup polymerization of isosorbide and a phosphonic dichloride. Isosorbide is a sucrose-determined diol, which has been found as a potential substitution for bisphenol-A [99]. An epic cyclic phosphonate monomer was created to produce water-solvent aliphatic poly (ethylenemethylphosphonate)s [100].

### 3.2.2 Poly[(organo)phosphazenes

There are two basic methods to synthesize polyphosphazene: first, ROP and second living cationic polymerization. Functioned polyphosphazene with a different kind of architecture can be prepared using controlled polymerization methods. Polydichlorophosphazene is extremely hydrolytically unstable but can be readily substituted with nucleophilic substitutes to give a wide range of stable polyorganophosphazenes with an extremely wide range of properties [101].

In the absence of water, the $[NPCl_2]_n$ undergoes through post-polymerization reaction with different organic nucleophiles. It is also termed as a macromolecular substitution reaction depending on the strength of nucleophile and steric hindrance [102].

It is necessary to ensure the complete postpolymerization with relatively high yield and purity; the unreacted chlorine group left in the polymer is a very important task. For this purpose, $^{31}P$ {$^1H$} NMR spectroscopy gives very valuable information for macromolecular substitution. From the NMR study, one can also conclude that in a certain macromolecular substitution reaction, substituent exchange reaction ($R^1$ can displace by $R^2$) also takes place to some extent. The cause of the substituent exchange reaction is different: steric hindrance and nucleophilicity of reagents (Figure 3.6) [103].

#### 3.2.2.1 Ring-opening polymerization of $[NPCl_2]_3$

Linear polyphosphazene ($NPCl_2$) can be synthesized from hexachlorocyclotriphosphazene, which is taken in a sealed glass tube under vacuum using ROP (Figure 3.5). The molecular weight of these polymers depends on reaction conditions such as temperature and reaction time. For better yield and high molecular weight, temperature and reaction time should be 250 °C and 18 h. At this condition, chlorine atoms present on cyclic trimmer $[NPCl_2]_3$ gets easily detached and makes cationic phosphazenium species. This cationic phosphazenium attacks at the second cyclic trimmer moiety and propagates the polymerization (Figure 3.7). The high molecular weights and broad polydispersities obtained from ROP is due to an initiation mechanism (cleavage of Cl from cyclic trimmer $[NPCl_2]_3$). When $CaSO_4.2H_2O$ as a promoter and $HSO_3(NH_2)$ as a catalyst are used, relatively reaction conditions (relatively lower temperature 210 °C and reaction time 6 h) reasonably change as comparing two

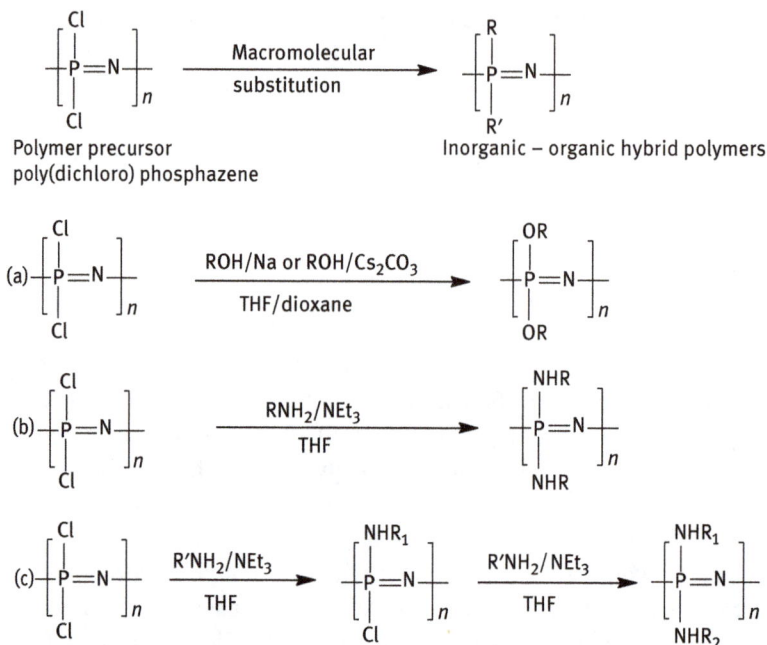

Polymer precursor
poly(dichloro) phosphazene

Inorganic – organic hybrid polymers

**Figure 3.6:** Synthesis of poly(organo)phosphazenes from the precursor poly(dichloro)phosphazene $[NPCl_2]_n$. The most common strategy to poly(organo)phosphazenes via macromolecular substitution using (a) alkoxides or aryloxides, (b) primary amines, and (c) mixed substitution.

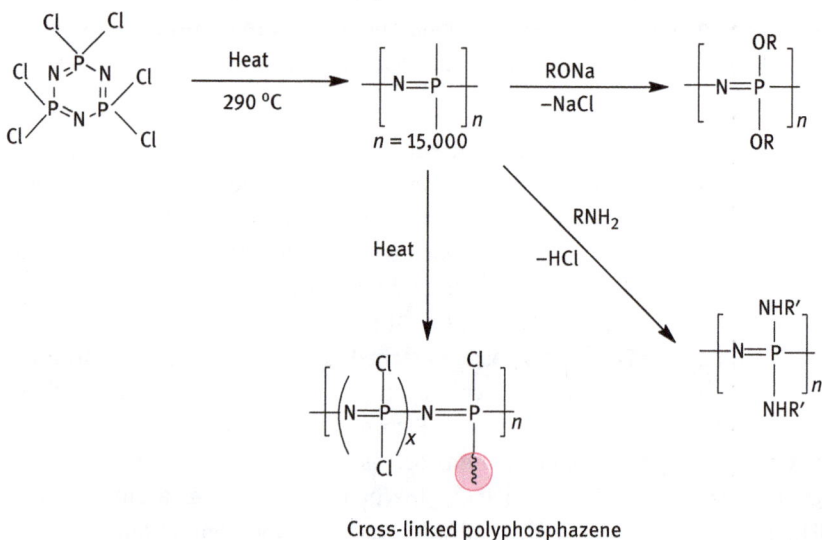

Cross-linked polyphosphazene

**Figure 3.7:** Synthesis of polyphosphazene from ring-opening polymerization and its substitution reaction.

aforementioned conditions [104]. Furthermore, phosphorus pentachloride ($PCl_5$) and ammonium chloride ($NH_4Cl$) may also be used for this purpose (Figures 3.7 and 3.8).

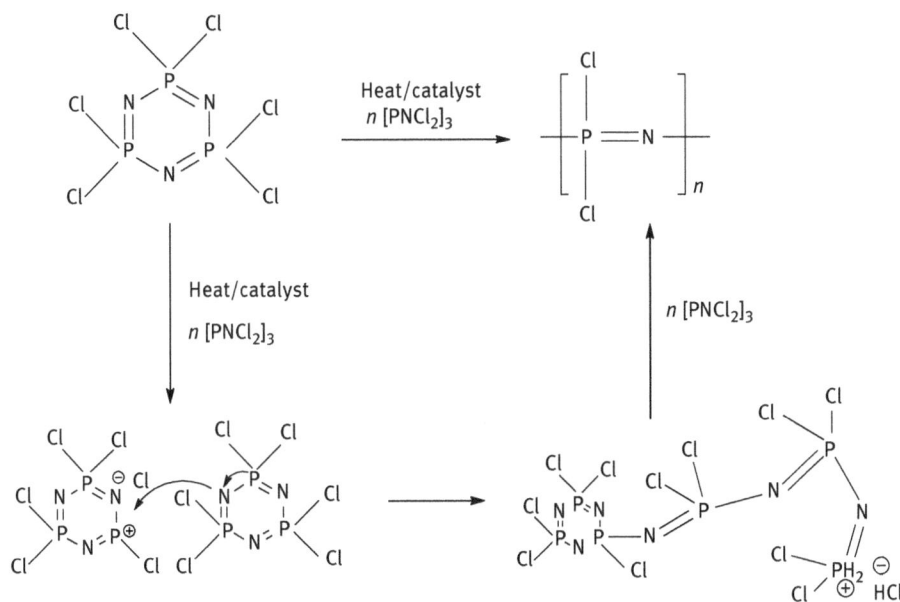

**Figure 3.8:** Thermally and/or catalytically induced ring-opening polymerization representing the classical route to synthesize poly(dichloro)phosphazene $[NPCl_2]_n$.

### 3.2.2.2 Controlled polymerization

Controlled radical polymerization is one of the very important methods to prepare bioconjugate, composites, and functioned copolymers for desired properties. The essential requirement for living polymerizations is that the reaction must have to follow first-order kinetics with respect to the concentration of monomer concentration. In each synthesis, definite molecular mass, narrow molecular mass distribution, and utilization of each end group are required. For that purpose, living cationic polymerization of trichlorophosphoranimine ($Cl_3PNSiMe_3$) can be employed to prepare polyphosphazene. The reaction of 1 mol of $Cl_3PNSiMe_3$ with 2 mol of $PCl_5$ gives a new cationic species $[Cl_3PNPCl_3]^+$ and $[PCl_6]^-$ at room temperature in specific solvent [105, 106].

### 3.2.2.3 One-pot facile polymerization

This technique is the very facile technique to prepare poly(organo) phosphazenes without using $[NPCl_2]_n$. Recently, Steinke and coworkers [107] have proposed a robust method for polymerization of $N$-silylphosphoranimines, initiated by $H_2O$ and catalyst $N$-methylimidazole. This method offers to synthesize a new range of polymers with a low polydispersity index (<1.15) and follow first-order polymerization kinetics. The simple, facile approach to synthesize polyphosphazene as a phosphate initiator is shown in Figure 3.9.

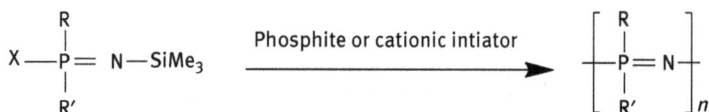

**Figure 3.9:** Direct synthesis route to poly(organo)phosphazenes.

### 3.2.2.4 Block copolymerization

The macromolecular substitution of [NPCl$_2$] can provide many copolymers with a random arrangement of monomers in the main backbone (Figure 3.10). The possible reason for these random copolymers is due to the nonselective chlorine atoms on the polyphosphazene backbone. As a result, many secondary monomer moieties can be used to form a block copolymer of polyphosphazene. Polynitrophosphazenes is also easily synthesized by simple nitration on polydiarylphosphazenes (Figure 3.11).

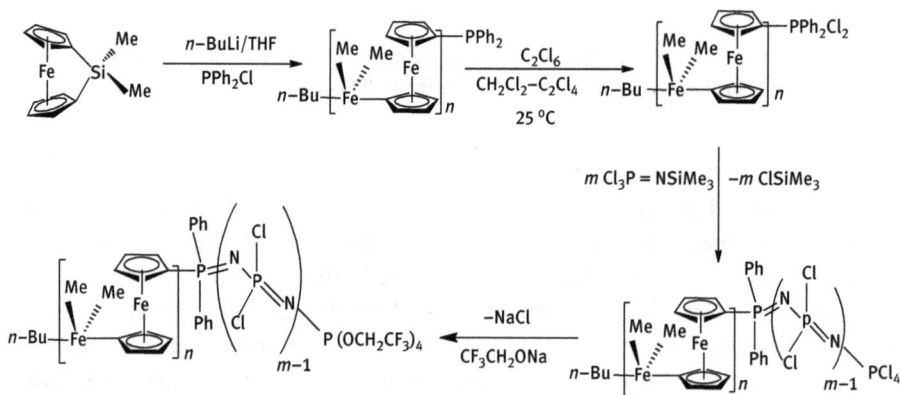

**Figure 3.10:** Synthesis of end-functionalized homopolymer and copolymer.

Soto and Manners [108] have synthesized block copolymer poly-(ferrocenylsilane-b-polyphosphazene) is prepared using the cationic polymerization of Cl$_3$PNSiMe$_3$ with macroinitiator diphenylphosphine groups as pendant groups on (poly (ferrocenylsilane)).

Telechelic polymers are those macromolecules whose further polymerization is possible using reactive sites of the side or end groups. Herein, polyphosphazene due to the macromolecular substitution of various functional groups onto the block

Poly(nitrophosphazenes)

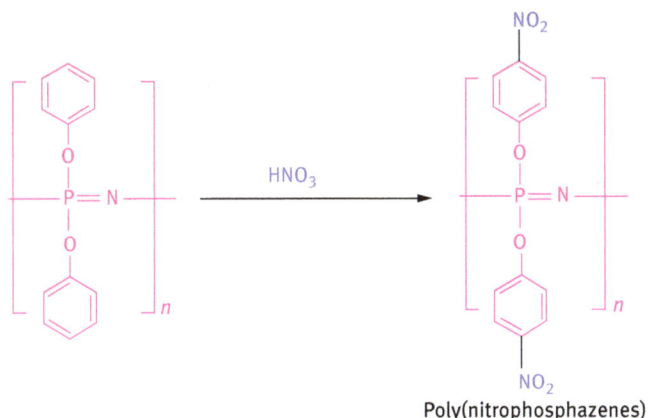

**Figure 3.11:** Synthesis of polynitrophosphazene by macromolecular substitution reaction.

copolymer may be possible to form new architecture with various properties. End cap polyphosphazene has a various polymerizable group, which can be employed to form different hyperbranched polymers [109–112].

### 3.2.3 Polyphosphoric acid

PPA responds with alcohols to deliver a phosphorylated item. This response is a great case of its amphoteric movement [113]. They examined that PPA responds with bitumen [114]. Giavarini et al. revealed that PPA responds with asphaltenes [115, 116]. Koebner and Robinson announced the utilization of PPA to create the skeleton of steroid [117]. PPA was utilized as dissolvable for various processes, for example, alkylation, acylation, cyclization, halogenation, drying out, hydrolysis, polymerization, and phosphorylation [118, 119]:

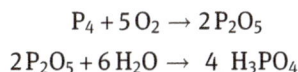

$$P_4 + 5\,O_2 \rightarrow 2\,P_2O_5$$
$$2\,P_2O_5 + 6\,H_2O \rightarrow 4\ H_3PO_4$$

**Scheme 1.** Reaction of phosphorus pentoxide

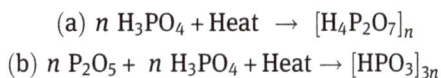

$$\text{(a) } n\ H_3PO_4 + \text{Heat} \rightarrow [H_4P_2O_7]_n$$
$$\text{(b) } n\ P_2O_5 +\ n\ H_3PO_4 + \text{Heat} \rightarrow [HPO_3]_{3n}$$

**Scheme 2.** Synthesis of PPA from different methods, that is, (a) dehydration and (b) dispersion methods. $n$ is an integer.

Polyphosphoric corrosive is created when phosphorus pentoxide ($P_2O_5$) responds with phosphoric corrosive ($H_3PO_4$) through the dry process [120]. Phosphorus is oxidized to phosphorus pentoxide, which takes shape as $P_4O_{10}$. Phosphorus pentoxide responds to water to deliver phosphoric corrosive.

Phosphoric corrosive ($H_3PO_4$) was delivered when sulfuric corrosive response with an apatite phosphate shake, that is, $Ca_3(PO_4)_2CaF_2$ through a wet process [121]. Unadulterated phosphoric corrosive was delivered either by the lack of hydration of $H_3PO_4$ at high temperatures or by warming $P_2O_5$ scattered in $H_3PO_4$ [122].

## 3.3 Properties

The glass transition temperature ($T_g$) is the unique properties of each and every polymer. It is responsible for the molecular flexibility. It is the temperature below which polymer becomes brittle or glassy and at this temperature, there is insufficient thermal energy to undergo significant tension dynamics. Above the glass transition temperature, the torsional motion is continuously increasing and an individual moiety can twist and yield to stress and strain. At this level, the polymer is a quasiliquid state (an elastomer) unless the bulk material has stiffened microcrystallite formation. Hence, one can conclude that a macromolecule with high value $T_g$ offers more resistance to bond torsion than a low valve $T_g$ macromolecule.

Polyphosphazenes have a very specific class of macromolecular science, which have very less, that is, negative value glass transition temperature, for example, $(NPCl_2)n = -66$ °C. $(NPF2)n = -96$ °C $[NP(OCH_3)_2]_n = -74$ °C; $[NP(OC_3H_7)_2]_n = -100$ °C [123, 124]. It can also be concluded that large and inflexible side groups can generate steric interference with each other, which is responsible for twisting motion.

The stiffness property of macromolecule may be due to the presence of microcrystalline domains (different arrangement of monomer moieties in the macromolecule) and after applying the temperature it will undergo liquid-like flow. Macromolecule with different pendant groups attached to the main chain usually lacks the symmetry because it does not control the stereoregularity of macromolecular systems. It is the only reason that polyphosphazenes with mixed substituent are amorphous in nature. Microcrystallinity in polymer science is observed in only those polymers that have different side groups such as fluoro, chloro, $CH_3$, $C_3H_7$, $OCH_2CF_3$, phenoxy, and various substituted phenoxy groups. Phosphazene with an amino functional group possesses both inter- and intramolecular hydrogen bonding; that is why they are amorphous in nature. Most aminophosphazene polymers are amorphous, perhaps because of intra- and intermolecular hydrogen bonding. The main chain of polyphosphazene is hydrophilic in nature, because of the presence of lone pair electrons on the nitrogen atoms of the constituent units and as a result, it can form hydrogen bonding to water molecules and at the same time hydrophobic nature is attributed to the presence of different organic groups present on the phosphorus atom of monomer moiety. As a result of both factors, its hydrophobic and hydrophilic properties can be decided.

Furthermore, cross-linking is taken place by the reaction of [NPCl$_2$] with a very small amount of water. Herein, this attack is due to the nucleophilic attack and the chlorine present on phosphorus atom can be easily replaced by an organic nucleophile; as a result, many polyphosphazenes with a wide range of properties can be formed. The bonding in polyphosphazene is unique as compared to other electron-rich organic system, neither unsaturated nor delocalized. P–N bond rotation is extremely low, but linear polyphosphazene is flexible experimentally, the only reason is due to different organic pendant groups. Pure phosphoric acid is a white crystalline solid having a melting point of 43.24 °C.

When phosphoric acid is diluted, it tastes sour and when phosphoric acid is concentrated, it is corrosive. It is a colorless, odorless, and viscous liquid when it is present in a less concentrated form. It is nontoxic and nonvolatile. Phosphoric corrosive has three acidic and replaceable hydrogen iotas. In this way, it responds contrastingly with various mineral acids. Phosphoric corrosive responds with the base to shape three classes of salts by the substitution of one, two, or three hydrogen molecules, for example, NaH$_2$PO$_4$, Na$_2$HPO$_4$, and Na$_3$PO$_4$ individually. At high temperatures, phosphoric corrosive particles can respond together and consolidate (with loss of water atomized) to form dimers, trimers, and even long polymeric chains, for example, PPAs and metaphosphoric acids:

$$2 \ H_3PO_4 \rightarrow H_4P_2O_7(-H_2O)$$

Phosphoric corrosive was a tribasic corrosive. There are three locales, focused on where the pH is equivalent to a p$K$ esteem, which is known as cradle areas. In the district centered around pH 4.7 (mid-path between the initial two p$K$ esteems) the dihydrogen phosphate particle, [H$_2$PO$_4$]$^-$, is the main species present.

In the area centered around pH 9.8 (mid-path between the second and third p$K$ esteems), the monohydrogen phosphate particle [HPO$_4$]$^{2-}$ is the main species present. This implies salts of the mono- and di-phosphate particles that can be specifically solidified from a fluid arrangement by setting the pH incentive to either 4.7 or 9.8. At the point when phosphorus corrosive is disintegrated in solid corrosive, antimony fluoride and phosphate hydroxide will be framed:

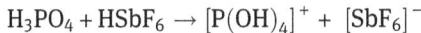

$$H_3PO_4 + HSbF_6 \rightarrow [P(OH)_4]^+ + [SbF_6]^-$$

The particle [P(OH)$_4$]$^+$ is isoelectronic with silicic corrosive, Si (OH)$_4$.

Phosphoric corrosive was created modernly when sulfuric corrosive response with apatite (tricalcium phosphate shake) through a wet procedure [125]:

$$Ca_5(PO_4)_3Cl + 5 \ H_2SO_4 + 10 \ H_2O \rightarrow 3 \ H_3PO_4 + 5 \ CaSO_4 \bullet 2 \ H_2O + HCl$$

Phosphoric corrosive was created mechanically when calcium hydroxyapatite responds with sulfuric corrosive [126]:

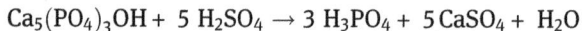

$$Ca_5(PO_4)_3OH + 5 H_2SO_4 \rightarrow 3 H_3PO_4 + 5 CaSO_4 + H_2O$$

The phosphoric corrosive arrangement contains 30% $P_2O_5$ and 40% $H_3PO_4$. This arrangement was concentrated to deliver business or vendor review phosphoric corrosive, which contains 50% $P_2O_5$ and 70% $H_3PO_4$. Further, the expulsion of water from the arrangement gives superphosphoric corrosive with 75% $P_2O_5$ and 100% $H_3PO_4$. Calcium sulfate (gypsum) is acquired as a result and expelled as phosphogypsum.

The phosphoric corrosive is filtered by evacuating mixes of arsenic and other possibly lethal pollutions. Phosphoric corrosive is readied when phosphorus (V) oxide breaks down in water. This technique gets readily unadulterated phosphoric corrosive.

Every single phosphoric corrosive is acidic as contrasted and mineral acids, for example, nitric corrosive [121]. As the chain length of phosphoric corrosive expands, the acidity of phosphoric corrosive additionally increments. Since each rehashed unit of phosphorus corrosive discharges a proton and balances out its charge by reverberation. The dimeric pyrophosphoric corrosive has two acidic hydrogens while the tremor triphosphoric corrosive has three acidic hydrogens [127].

## 3.4 Application of polyphosphate

### 3.4.1 Anticorrosive

The high alloy steel shows excellent resistivity toward corrosion. Steel shows passiveness even in the acidic solution like industrial phosphoric acid and simplified solution of pure $H_3PO_4$. But the presence of chemical impurities as $Cl^-$ and $F^-$ in acidic media leads to a decrease in resistivity and increases metal dissolution, while using industrial phosphoric acid efficiency to inhibit corrosion enhances because of the formation of the polyphosphate film on the stainless steel surface that prohibits the aggressive ions to approach the surface. The polyphosphate helps promote higher crystallization of $Cr_2O_3$ in the inner oxide film, which provides higher resistance to corrosion via a lesser amount of oxygen and/or cationic vacancies. Furthermore, if calcium and sulfate are present, then they get incorporated in the polyphosphate film, which increases the thickness of the film, thereby hindering the transport of the aggressive ions to the passive film [128]. Cyclohexaphosphate made up of two alkaline and bivalent cations can be coordinated with a metal polyhedral to give a high-dimensional architecture. The electrode coated with this material could be regarded as promoting the polyphosphate thin film, which would serve as anticorrosive protection and in batteries [129].

### 3.4.2 Flame retardant

Many compounds used in daily life catch fire easily. Therefore, to make them more fire resistant they are reacted with a negatively charged ammonium polyp. Layer-to-layer coating of cotton by mild reaction between APP and branched polyethylenei-mine (PEI) provides an ecofriendly, cost-effective, fast, and wash-durable, fire-resistant coating for the cellulose-rich cotton fabrics [130]. The combination of short wool fibers and polypropylene prepared by melt compounding followed by compression molding has more tensile strength than a composite of polypropylene, wool, and APP. Basically, APP must have a positive effect on mechanical strength, but it is not so because of deteriorated compatibility between wool fibers and polypropylene, the reinforcing effect of wool in the composites still induced higher tensile strengths than polypropylene and APP composite. The composite sample exhibits fire retarding properties and also slows down the polymer dripping phenomenon [131]. APP modified by inorganic chemical precipitation reaction that was carried out on the surface area of APP has high thermal stability, good water repelling, and compatibility with polyphosphate matrix [132].

Cotton along with synthetic fiber such as nylon has softness, durability, and strength. Both the components are flammable. They cannot be even protected by a complex of PEI and a polyphosphate. Whereas the aqueous complex of PEI and APP fabricates melamine polyphosphate. Polyphosphate was found to be effective in total reduction of heat [133]. Melamine foam with two double layers covering chitosan (positively charged) and APP (positively charged) has excellent properties like self-extinguishing, fire stable, less shrinking, and less heat release. Hence, the use of waterborne, environmentally benign components provides environmentally friendly flame-retardant treatment for melamine foam [134]. The combination poly(4,6-dichloro-$N$-butyl-1,3,5-triazin-2-amine-ethylenediamine) as a charring agent and APP suppresses the fire and smoke created on combustion of polypropylene because char has P–O–P and P–O–C cross-linking structures and polyaromatic structures. Thus, it leads to good thermal stability and char ability [135]. $N$-alkoxy hindered amine (NOR116) has free radical quenching property. Its synergistic effect with APP/pentaerythritol helps in increasing fire, ultraviolet age resistance, and thermal degradation of polypropylene. The role of NOR116 is to inhibit the chain reaction in the gas phase and accelerate the generation of thermo-stable Intumescent char in the condensed phase [136]. The results reveal that the carbon and nitrogen element contents of the char layer of polyphosphate/IFR (Intumescent flame retardant)/NOR116 are much more as compared to only polyphosphate/IFR.

### 3.4.3 Composites

Epoxy resin is one of the most common thermosetting resins, which is widely used as adhesives, coatings, laminates, and electronic/electrical appliances because of its low cost and outstanding performance [137, 138]. Even after such promising property, it possesses high flammability that restricted its utility in electronic applications [139]. Therefore, it is important to make it the flame-retardant grade. Zhao et al. [140] studied the synthesis of poly (4,40-domino diphenyl sulfone 2,6,7-trioxa-1-phospha-bicyclo[2.2.2]octane-4-methanol-substituted phosphoramide) (PSA) by the solution polycondensation method. The PSA alone or in the combination of APP provides positively the flame retardancy and thermal stability of epoxy resins. It also reduces the peak heat release rate, total heat release, and total smoke release due to PSA. Whereas the increase in the amount of residual char and decrease in the number of pyrolysis products during combustion were observed due to APP as its form of aromatic structures bridged by P–O–P and P–O–C bonds.

Similarly, APP in epoxy resin modified by using hyperbranched PEI by cation exchange reaction results in flame retardance, thermal stability, less heat release, and smoke suppression. Because in such case, an increase in temperature generates P–N–C group, which further decompose into N-containing nonflammable gases. Thus, forming residue with P–N–C, P–O–P, and C=C structures provides resistivity toward heat, oxygen, and combustion. This polyamine hardener provides a wide range of application of epoxy resin [141].

The alkaline lignin as additives provides intercalates with both b-sheets and a-helix structures and APP increases charring effect on composites based on thermoplasticized Zein [142]. APP montmorillonite nanocomposite and OPS/9,10-dihydro-9-oxa-10-phosphaphenanthrene-10-oxide make epoxy resins fire inert by the formation of –Si-O-C- or –Si-O-(P = O)-C-cross-linked structures in the condensed phase under the action of heat, resulting in solid carbonaceous char [143]. The combination of nanosilica and nanoalumina along with APP in polymethyl methacrylate (PMMA) exhibits formation of a very cohesive and expanded a layer containing silicon pyrophosphate. Because of APP, the introduction of nanosilica on the surface also reduces the flammability and leads to the delay of the CO yield emission, which could allow people to evacuate buildings or houses in case of fire [144].

Ethylene-vinyl acetate copolymer (EVA) is an insulator used in cable coating. Properties like low toxicity, low smoke, halogen free, and high efficiency can be provided by IFR [145], even though this strategy has limitations of weak water resistance and poor compatibility with the polymer matrix. Therefore, Sheng et al. [146] reported microencapsulation of APP with polyester polyurethane decreasing the solubility of EVA in water. Whereas on irradiation with 180 kGy microencapsulated ammonium polyphosphate (MCAPP) gets immobilized in a 3D cross-linked structure at a higher extent than irradiated with 120 kGy resulting in an increase in the flame resistance and electrical insulation properties of the EVA composites. While carrying out hot water test at 80 °C for around 2 weeks, it was observed that the composite exhibits same performance with 180 kGy, which shows that microencapsulation helps to obtain stability and lower electrical fire hazard risk due to higher thermal stability while carrying out long-term hot water aging test.

Flame-retarded filler polyphosphate enhances the hydrolysis of polylactic acid [147]. Polylactic acid flammability can also be reduced by the incorporation of APP-based FR additive and montmorillonite clays in a weight ratio of 10:1 into the matrix layers. Introduction of APP shows a positive change in various properties like [148]

- mechanical properties,
- the stiffness of their matrix layers,
- fiber–matrix bonding, and
- energy absorption capacity (impact perforation energy as high as 16 J/mm).

As FR-free polymer fiber pulls out first, whereas in 10 wt% FR containing poly (lactic acid)-self reinforced composite (PLA-SRC) matrix is broken first. On the contrary, 16 wt% FR containing PLA-SRC shows both; the matrix and the fibers undergo about the same strain (35%) during deformation indicating strong bonding.

Spent coffee ground is one of the daily wastes, but this waste can be used for the production of P, N, co-doped C on reacting with APP. Ratios of spent coffee grounds to APP (1:0.2, 1:0.4, and 1:0.8) influence the surface area and specific capacitance value. Sample with higher APP showed the highest surface area (999.64 $m^2/g$) with the presence of micropores, mesopores, macropores, the existence of a high concentration of dopants (N, P, and O), and high pore volume. Also, it showed high specific capacitance values (286 F/g in 1 M $H_2SO_4$) in acid and alkaline conditions [149].

### 3.4.4 Biomedical

In medical science, there is a need for appropriate prefabricated bone graft materials, because of the shortage and risks of autogenic and allogeneic implants. Therefore, sodium-based polyphosphate after calculation by calcium chloride form calcium polyphosphate, which has a potential scaffold for bone implants [150].

Dental hypersensitivity is a common and costly disease; it can be resolved by using amorphous microparticles, prepared from the natural polymer (polyphosphate). It is used to reseal the dentinal tubules exposed and reduce by that the hypersensitivity [151]. Calcium bonded with polyp forms a nanosphere, which shows the morphogenetic activity. The nanoparticle encapsulating retinol shows a synergistic effect on the expression of all fibrillar collagen types studied (types I, II, and III) [152]. Erosion of teeth is a common problem these days. It can be cured to some extent by the use of fluoride alone or in combination with polyphosphate as a complementary preventive measure. do-Amaral et al. [153] studied the decrease in the rate of dissolution of hydroxyapatite in terms of caries and erosion by polyphosphate salts, sodium trimetaphosphate (TMP) and SHMP associated or not with fluoride (F).

Polyphosphate is also found in eukaryotic organisms to carry out the biological function. It is deposited in tissue or organelles but it is an insoluble polymer. In order to make it useful for the biological purpose, it has to be soluble. Wang et al. [1] studied that inorganic polyphosphate (poly P) is a physiological polymer composed of tens to hundreds of phosphate units linked together via phosphoanhydride bonds. It has the ability to induce bone differentiation in osteoblasts [154]. Anterior eye also has diadenosine polyphosphates such as Ap4A that acts at P2 receptors, which leads to modulation of the rate and composition of tear secretion, impact corneal wound healing, and lower IOP [155].

Rodrigues et al. [156] prepared a hybrid of chitosan, carrageenan, tripolyphosphate nanoparticles as protein carrier and studied its application in pulmonary and nasal transmucosal delivery of macromolecules. The role of tripolyphosphate is to decrease the size and stabilization of nanoparticles. In this bovine serum albumin (BSA) undergoes microencapsulation by nanoparticles by spraydrying to get the aerodynamic requirements inherent to pulmonary delivery. Vanillin and TPPP cross-linker along with chitosan were also used to prepare microspores that are encapsulated by placebo (C1), BSA (C2), monovalent tetanus toxoid (TT) (C3), divalent tetanus (TT), and diphtheria toxoids (DT) (C4). Encapsulated microspore had amorphous nature, had more surface area, and was thermally stable. It has medical use as a multivalent oral vaccine [157].

Sodium tripolyphosphate is a nontoxic inorganic polyanion. The chitosan and sodium tripolyphosphate reacted with ionic chelation process to synthesize chitosan nanoparticles. Methyl-4-(2-chloro-4-fluorophenyl)-2-(3,5-difluoro-2-pyridinyl)-6-methyl-1,4-dihydro-pyrimidine-5-corboxylate (Bay41-4109) has poor aqueous solubility. Therefore, it becomes a challenge to use it as an oral drug, and coating it with chitosan and sodium tripolyphosphate make it soluble. As nanoparticles act as carriers for the hydrophobic Bay41-4109, it improves oral drug delivery by enhancing the bioavailability and dissolution rates of hydrophobic drugs [158].

Tripolyphosphate has been used as a cross-linker [159] for interlocking chitosan hydrogel beads in the synthesis of silver nanoparticles. The nanocomposite has effectively used in drug delivery, swelling, and antibacterial effect toward *Escherichia coli* and *Staphylococcus aureus* bacteria [160]. The cross-linked property of sodium TMP was observed in the fabrication of microgel of carboxymethyl cellulose polymers. Sodium TMP ($Na_3P_3O_9$) attacks on the hydroxyl (-OH) groups of carboxymethyl starch, generating the formation of ester linkages [161]. The protein uptake of lysozyme–microgel complex, that is, uptake of lysozyme by microgel, decreases with increases in cross-linking density. The absorption of lysozyme with the help of microbes showed that the protein uptake increased with rising pH and concentration strength of lysozyme [162].

Nontoxic reagent sodium TMP has been used for synthesis of β-cyclodextrin/ dextran-based hydrogen. Its swelling property depends on the content of phosphate group, which provides electrolytic nature to the hydrogel. The increase in swelling occurs with an increase of cyclodextrin/dextran and a decrease in phosphate-based cross-linkers. The releasing and loading capacity properties for methylene blue and benzophenone exhibited their application for drug delivery [163].

Liu et al. [164] prepared redox-responsive, good biocompatibility, and biodegradability polyphosphate micelles and exhibited its applicability reducing cell proliferation by delivering anticancer drugs into the nuclei of tumor cells as denoted in Figure 3.12.

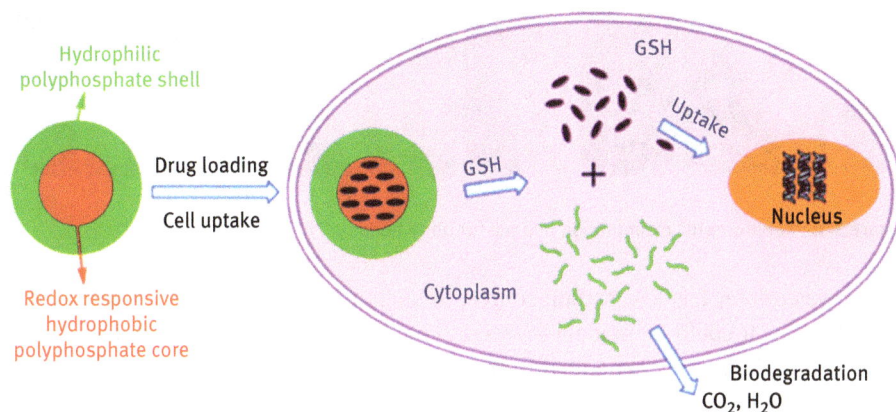

**Figure 3.12:** Scheme of delivering anticancer drugs [164].

Li et al. [165] synthesized biocompatible and biodegradable Cu-doped calcium polyphosphate scaffolds combined with Cu preconditioned bone marrow mesenchymal stem cells to treat bone defects that could enhance new bone formation and defect healing as shown in Figure 3.13.

## 3.4.5 Remedies of environmental pollutants

Phosphorus-accumulating organisms (PAOs) such as *Candidatus Accumulibacter phosphatis* Clade I and II take up and store phosphate as an intercellular polyphosphate. Therefore, PAO is regarded as a tool for the elimination of enhanced biological phosphorus removal (EBPR) from wastewater [166]. The wastewater has polyphosphorus, which could be eliminated by EBPR [167]. The biological removal of phosphate can be carried out with polyphosphate accumulating organisms (PAOs) [168]. Thus, PAOs such as *Candidatus Accumulibacter* [169], Comamonadaceae [170] in wastewater have low dissolved oxygen and low solid retention time conditions. The

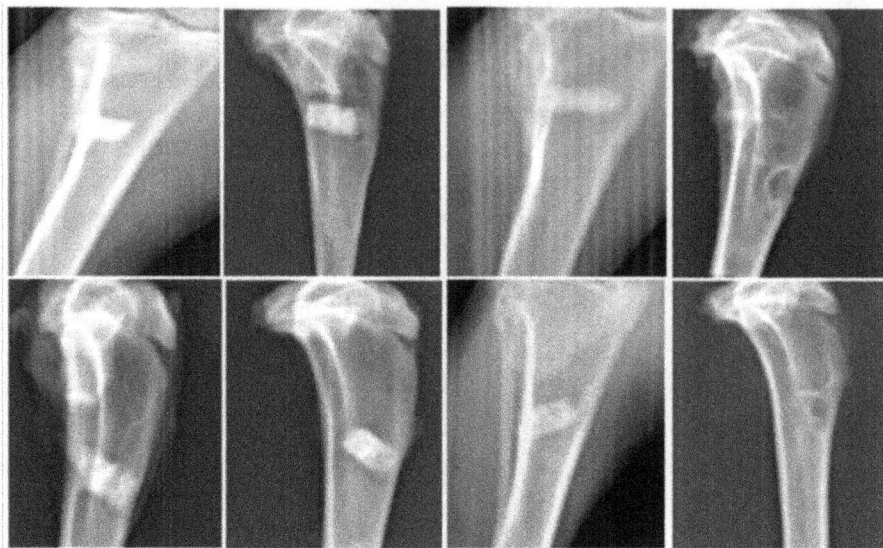

**Figure 3.13:** Radiographic examination images for bone healing [165].

urban wastewater has less carbon source; in such a case, this method becomes less cost efficient. It could be made a low-cost process by using crude glycerol, which is an industrial byproduct generated in the biodiesel production. Crude glycerol along is used along with methanol, salts, VFA, and long-chain fatty acids (LCFA). In this VFA and methanol show increased PAO activity, whereas long-term operation of LCFA gets failed because of increased hydrophobicity of the sludge [171].

Wildfires are an important issue, which can be solved by using fire and fire-fighting chemicals such as APP. It has the largest and longest-lasting impacts on soil plant systems even after 10 years. APP introduction also helps in increasing soil and plant P levels, and Na/K ratio in leguminous shrubs [172].

### 3.4.6 Food industries

Polyphosphate acts as a functional additive for enhancing protein gelatin [173]. The traditional sausages named as salpicao are prepared by fermentation of raw pork meat. Sometimes, the formation may lead to the generation of Enterobacteriaceae, *Staphylococcus aureus*, and *Listeria monocytogenes*, which make it pathogenic and is therefore not preferred as an edible item. Polyphosphate helps in reduction of fermentation by making it less favorable for bacterial growth [174].

Protein products like meat produced from animals have fat and sodium. For good quality of the protein, it is required to reduce both these elements. Its direct reduction by technologies is very tedious and problematic [175]. While consuming such protein, sodium percentage increases in the body by taking sodium salt in many foodstuffs [176]. Therefore, replacement of NaCl by KCl and sodium tripolyphosphate on low-fat meat sausages formulated with fish oil was studied by Marchetti et al. [177]. These changes only affected the matrix microstructure, and not on sensory acceptability.

Fusarium wilt of tomato caused by *Fusarium oxysporum* f. sp. *lycopersici* (Sacc.) is highly destructive to tomatoes grown in glasshouse and in fields across the world. The cross-link of cell wall chitosan with protecting plants from this disease by the delay in wilt disease symptom expression and reduce the wilt disease severity [178].

### 3.4.7 Mineral dissolution in the oral cavity

PolyP accumulation is well known for its association with certain microbes to resist physical and chemical stresses, and also provides an alternative source of energy in different environmental conditions [179, 180]. It has been demonstrated that polyaccumulating bacteria (PAB) are capable of modulating the ionic constituents in equilibrium with apatite group minerals in pore waters and subsequently altering the saturation state of the surrounding fluids resulting in microenvironments that are thermodynamically favorable for mineral precipitation [181, 182]. Further study of metabolic processes of PAB has achieved a new paradigm in our understanding of the modulation of $PO_4^{3-}$ and $Ca^{2+}$ activities and their relationship to the solubility of calcium phosphate minerals. Although localized acid production in cryogenic biofilms undoubtedly impacts mineral solubility, biological influence on the chemical saturation of $Ca^{2+}$ and $PO_4^{3-}$ may present an additional component to the development and rapid progression of carious lesions. In order to address several different facets of the hypothesis that PAB may affect localized chemical saturation in the oral cavity, analysis of genomic databases of oral taxa and quantity of PAB in clinical samples of plaque, saliva, and dentinal lesions perform phosphate uptake experiments using a defined in vitro single-species model, and find out the potential impact of polyP accumulation on the saturation state of saliva [183].

The sequestration of $PO_4^{3-}$ by PAB has the potential to alter the chemical conditions of the oral environment, which promote mineral dissolution under certain conditions in the mouth, leading to dental decay. Alternatively, the concentrated release of $PO_4^{3-}$ from PAB could lead to the precipitation of dental calculus (mineralized dental plaque) under a different set of oral microenvironmental conditions. These ions can also be incorporated into various other phases of apatite such as fluorapatite ($Ca_5(PO_4)_3F$) and carbonate-hydroxyapatite ($Ca_5(PO_4, CO_3)_3(OH)$). These substitutions are common in the oral cavity and vary from individual to individual, as well as from tooth to tooth. Mineral solubility may increase or decrease depending on the substitutions in the lattice structure [184]. In order to assess the controls on PAB metabolisms and their potential roles in altering the saturation chemistry of the saliva/mineral interface, a comprehensive understanding of the ecology and physiology of PAB in the oral environment is needed [185]. Clinical assessments and in situ taxonomic identification of PAB in oral biofilms will aid us in understanding their ecophysiologies, as well as aid in our ability to treat oral diseases such as dental caries that remain incompletely understood.

## 3.5 Conclusion

Polyphosphate is referred to as a salt or ester of polymeric oxyanion bonded by sharing of oxygen atom from tetrahedral phosphate structure units. It may have a linear or cyclic structure. Each unit of high-polymeric inorganic polyphosphates is bonded by high-energy phosphoanhydride bonds. Previously, phosphorus is considered as a "molecular fossil" as it is used as an energy source by microorganisms in adverse situation. In prebiotic evolution, the polyp is found in bacterial species as conserved enzyme PPK encoded by ppk gene. It is reported that the principal enzyme in many bacteria is responsible for the synthesis of inorganic polyphosphate (poly P) from ATP. Later, it is observed that it plays a regulatory role in all living organisms, and the metabolic activity plays a major role in the genetic and enzymatic level. It is also naturally occurring in soil, water bodies' sediment, rocks, and others. The presence plays a crucial role of abiotic to biotic component, that is, it is essential for the growth of the plant, which is approved as an additive for food, treating wastewater, reducing dental problems, drug delivery, and so on. Addition of polyphosphate and its derivative increases flame retardancy, thermal stability, mechanical, and tensile strength in various products such as wood, paper, fabric, and plastic. Polyphosphate also functions as a commercial retardant for forest fires and via the role its ammonium content plays in forming a protective charred layer after burning, thereby preventing further ignition.

Polyphosphonate and polyphosphazene are also very greatly diverse performance materials with greater radiation resistance, high refractive index, ultraviolet and visible transparency, and also fire resistant.

## References

[1]   Wang, X., Huang, J., Wang, K., Neufurth, M., Schroder, H.C., Wang, S., & Muller, W.E.G. The morphogenetically active polymer, inorganic polyphosphate complexed with GdCl$_3$, as an inducer of hydroxyapatite formation in vitro, Biochem. Pharmacol., 2016, 102, 97–106.

[2]   Tanaka, K., Fujiya, M., Konishi, H., Ueno, N., Kashima, S., Sasajima, J., Moriichi, K., Ikuta, K., Tanabe, H., & Kohgo, Y. Probiotic-derived polyphosphate improves the intestinal barrier function through the caveolin-dependent endocytic pathway, Biochem. Biophys. Res. Commun., 2015, 467, 541–548.

[3]   Dale, A.W., Boyle, R.A., Lenton, T.M., Ingall, E.D., & Wallmann, K. A model for microbial phosphorus cycling in bioturbated marine sediments: Significance for phosphorus burial in the early Paleozoic. Geochim. Cosmochim. Acta., 2016, 189, 251–268.

[4]   Jorgensen, C., Inglett, K.S., Jensen, H.S., Reitzel, K., & Reddy, K.R. Characterization of biogenic phosphorus in outflow water from constructed wetlands, Geoderma, 2015, 257, 58–66.

[5]   Park, T., Ampunan, V., Lee, S., & Chung, E. Chemical behavior of different species of phosphorus in coagulation, Chemosphere, 2016, 144, 2264–2269.

[6]   Cade-Menun, B.J. Improved peak identification in 31P-NMR spectra of environmental samples with a standardized method and peak library, Geoderma, 2015, 257, 102–114.

[7]   Stutter, M.I., Shand, C.A., George, T.S., Blackwell, M.S.A., Dixon, L., Bol, R., MacKay, R.L., Richardson, A.E., Condron, L.M., & Haygarth, P.M. Land use and soil factors affecting accumulation of phosphorus species in temperate soils, In Special Issue on Developments in Soil Organic Phosphorus Cycling in Natural and Agricultural Ecosystems. Geoderma, 2015, 257, 29–39.

[8]   Zhang, Y., He, J., & Yang, R. The effects of phosphorus-based flame retardants and octaphenyl polyhedral oligomeric silsesquioxane on the ablative and flame-retardation properties of room temperature vulcanized silicone rubber insulating composites, Polym. Degrad. Stab., 2016, 125, 140–147.

[9]   Leistner, M., Abu-Odeh, A.A., Rohmer, S.C., & Grunlan, J.C. Water based chitosan/melamine polyphosphate multilayer nanocoating that extinguishes fire on polyester-cotton fabric, Carbohydrate Polym., 2015, 130, 227–232.

[10]  Vahabi, H., Lin, Q., Vagner, C., Cochez, M., Ferriol, M., & Laheu, P. Investigation of thermal stability and flammability of poly(methyl methacrylate) composites by the combination of APP with $ZrO_2$, sepiolite or MMT, Polym. Degrad. Stab., 2016, 124, 60–67.

[11]  Schirp, A., & Su, S. Effectiveness of pre-treated wood particles and halogen-free flame retardants used in wood-plastic composites, Polym. Degrad. Stab., 2016, 126, 81–92.

[12]  Feng, C., Zhang, Y., Liang, D., Liu, S., Chi, Z., & Xu, J. Influence of zinc borate on the flame retardancy and thermal stability of intumescent flame retardant polypropylene composites, J. Analyt. Appl. Pyrol., 2015, 115, 224–232.

[13]  Chen, C., Gu, X., Jin, X., Sun, J., & Zhang, S. The effect of chitosan on the flammability and thermal stability of polylactic acid/ammonium polyphosphate biocomposites, Carbohydr. Polym., 2017, 157, 1586–1593.

[14]  Carosio, F., & Alongi, J. Influence of layer by layer coatings containing octa propyl ammonium polyhedral oligomeric silsesquioxane and ammonium polyphosphate on the thermal stability and flammability of acrylic fabrics, J. Analyt. Appl. Pyrol., 2016, 119, 114–123.

[15]  Penczek, S., Pretula, J., Kubisa, P., Kaluzynski, K., & Szymanski, R. Reactions of $H_3PO_4$ forming polymers. Apparently simple reactions leading to sophisticated structures and applications, Prog. Polym. Sci., 2015, 45, 44–70.

[16]  Bae, W.J., Jue, S.S., Kim, S.Y., Moon, J.H., & Kim, E.C. Effects of sodium tri- and hexametaphosphate on proliferation, differentiation and angiogenic potential of human dental pulp cells, J. Endod., 2015, 41, 896–902.

[17]  Kornberg, A. Inorganic polyphosphate: Toward making a forgotten polymer unforgettable, J. Bacteriol., 1995, 177, 491–496.

[18]  Rao, N.N., Gómez-García, M.R., & Kornberg, A. Inorganic polyphosphate: Essential for growth and survival, Annu. Rev. Biochem., 2009, 78, 605–647.

[19]  Rashid, M.H., Rao, N.N., & Kornberg, A. Inorganic polyphosphate is required for motility of bacterial pathogens, J. Bacteriol., 2000, 182, 225–227.

[20]  Boetsch, C., Aguayo-Villegas, D.R., Gonzalez-Nilo, F.D., Lisa, A.T., & Beassoni, P.R. Putative binding mode of Escherichia coli exopolyphosphatase and polyphosphates based on a hybrid in silico/biochemical approach, Archiv. Biochem. Biophys., 2016, 606, 64–72.

[21]  Shiba, T., Tsutsumi, K., Ishige, K., & Noguchi, T. Inorganic polyphosphate and polyphosphate kinase: Their novel biological functions and applications, Biochem. (Mosc), 2000, 65, 315–323.

[22] Kim, K.S., Rao, N.N., Fraley, C.D., & Kornberg, A. Inorganic polyphosphate is essential for long-term survival and virulence factors in Shigella and Salmonella spp, Proc. Natl. Acad. Sci. U S A, 2002, 99, 7675–7680.

[23] Kornberg, A., Rao, N.N., & Ault-Riché, D. Inorganic polyphosphate: A molecule of many functions, Annu. Rev. Biochem., 1999, 68, 89–125.

[24] Kornberg, A. Inorganic polyphosphate: A molecule of many functions, Prog. Mol. Subcell. Biol., 1999, 23, 1–18.

[25] Angelova, P.R., Agrawalla, B.K., Elustondo, P.A., Gordon, J., Shiba, T., Abramov, A.Y., Chang, Y.T., & Pavlov, E.V. In situ investigation of mammalian inorganic polyphosphate localization using novel selective fluorescent probes JC-D7 and JC-D8, ACS Chem. Biol., 2014, 9, 2101–2110.

[26] Golebiewska, E.M., & Poole, A.W. Platelet secretion: From haemostasis to wound healing and beyond, Blood Rev., 2015, 29, 153–162.

[27] Montilla, M., Hernández-Ruiz, L., García-Cozar, F.J., Alvarez-Laderas, I., Rodríguez-Martorell, J., & Ruiz, F.A. Polyphosphate binds to human von Willebrand factor in vivo and modulates its interaction with glycoprotein Ib, J. Thromb. Haemost., 2012, 10, 2315–2323.

[28] Geddings, J.E., & Mackman, N. New players in haemostasis and thrombosis, J. Thromb. Haemost., 2014, 111, 570–574.

[29] Morrissey, J.H., & Smith, S.A. Polyphosphate as modulator hemostasis, thrombosis, and inflammation, J. Thromb. Haemost., 2015, 13, S92–S97.

[30] Santi, M.J., Montilla, M., Carroza, A., & Ruiz, F.A. Novel assay for prothrombotic polyphosphates in plasma reveals their correlation with obesity, Thromb. Res., 2016, 144, 53–55.

[31] Plug, T., & Meijers, J.C.M. Stimulation of thrombin- and plasmin-mediated activation of thrombin-activatable fibrinolysis inhibitor by anionic molecules, Thromb. Res., 2016, 146, 7–14.

[32] Stewart, R.J., Wang, C.S., Song, I.T., & Jones, J.P. The role of coacervation and phase transitions in the sandcastle worm adhesive system, Adv. Colloid. Interface Sci., 2016, 239, 88–96.

[33] Sánchez-Thomas, R., Moreno-Sánchez, R., & García-García, G.D. Accumulation of zinc protects against cadmium stress in photosynthetic Euglena gracilis, Environ. Exp. Bot., 2016, 131, 19–31.

[34] Jiang, L.G., Liang, B., Xue, Q., & Yin, C.W. Characterization of phosphorus leaching from phosphate waste rock in the Xiang xi River watershed, three gorges reservoir, China Chemosphere, 2016, 150, 130–138.

[35] Wan, S., Tieu, A.K., Xia, Y., Zhu, H., Trans, B.H., & Cui, S. An overview of inorganic polymer as potential lubricant additive for high temperature, Tribology Inter., 2016, 102, 620–635.

[36] Sharma, V., Erdimer, A., & Aswath, P.B. An analytical study of tribofilms generated by the interaction of ashless antiwear additives with ZDDP using XANES and nano-indentation, Tribology Inter., 2015, 82, 43–57.

[37] Sharma, V., Gable, C., Doerr, N., & Aswath, P.B. Mechanism of tribofilm formation with P and S containing ionic liquids, Tribology Inter., 2015, 92, 353–364.

[38] Steinbach, T., & Wurm, F.R. Poly(phosphoester)s: A new platform for degradable polymers, Angew. Chem., Int. Ed., 2015, 54, 6098–6108.

[39] Bauer, K.N., Tee, H.T., Velencoso, M.M., & Wurm, F.R. Main chain poly(phosphoester)s: History, synthesis, degradation, bio-and flame-retardant applications, Prog. Polym. Sci., 2017, 73, 61–122.

[40] Du, J.Z., Du, X.J., Mao, C.Q., & Wang, J. Tailor-made dual pH sensitive polymer–doxorubicin nanoparticles for efficient anticancer drug delivery, J. Am. Chem. Soc., 2011, 133, 17560–17563.

[41] Mao, C.Q., Du, J.Z., Sun, T.M., Yao, Y.D., Zhang, P.Z., Song, E.W., & Wang, J. A biodegradable amphiphilic and cationic triblock copolymer for the delivery of siRNA targeting the acid ceramidase gene for cancer therapy, Biomaterials, 2011, 32, 3124–3133.

[42]  Sun, T.M., Du, J.Z., Yao, Y.D., Mao, C.Q., Dou, S., Huang, S.Y., Zhang, P.Z., Leong, K.W.;.,
      Song, E.W., & Wang, J. Simultaneous delivery of siRNA and paclitaxel via a "two-in-one"
      micelleplex promotes synergistic tumor suppression, ACS. Nano., 2011, 5, 1483–1494.

[43]  Schöttler, S., Becker, G., Winzen, S., Steinbach, T., Mohr, K., Landfester, K., Mailänder, V., &
      Wurm, F.R. Protein adsorption is required for stealth effect of poly (ethylene glycol)-and poly
      (phosphoester)-coated nanocarriers, Nat. Nanotechnol., 2016, 11, 372–377.

[44]  Müller, J., Bauer, K.N., Prozeller, D., Simon, J., Mailänder, V., Wurm, F.R., Winzen, S., &
      Landfester, K. Coating nanoparticles with tunable surfactants facilitates control over the
      protein corona, Biomaterials, 2017, 115, 1–8.

[45]  Steinmann, M., Wagner, M., & Wurm, F.R. Poly-(phosphorodiamidate)s by olefin metathesis
      polymerization with precise degradation, Chem. Eur. J., 2016, 22, 17329–17338.

[46]  Wang, H., Su, L., Li, R., Zhang, S., Fan, J., Zhang, F., Nguyen, T.P., & Wooley, K.L.
      Polyphosphoramidates that undergo acid-triggered backbone degradation, ACS. Macro.
      Lett., 2017, 6, 219–223.

[47]  Wolf, T., Steinbach, T., & Wurm, F.R. A library of well-defined and water-soluble poly(alkyl
      phosphonate)s with adjustable hydrolysis, Macromol., 2015, 48, 3853–3863.

[48]  Horsman, G.P., & Zechel, D.L. Phosphonate biochemistry, Chem. Rev., 2017, 117, 5704–5783.

[49]  Ogawa, T., Nushimatsu, T., & Minoura, Y. Condensation polymerization of
      p-phenylphosphonic dichloride with diamines, Macromol. Chem., 1968, 114, 275–283.

[50]  Pretula, J., Kaluzynski, K., Szymanski, R., & Penczek, S. Preparation of poly(alkylene
      h-phosphonate)s and their derivatives by polycondensation of diphenyl h-phosphonate with
      diols and subsequent transformations, Macromol., 1997, 30, 8172–8176.

[51]  Pretula, J., Kaluzynski, K., Szymanski, R., & Penczek, S. Transesterification of oligomeric
      dialkyl phosphonates, leading to the high-molecular-weight poly-H-phosphonates, J. Polym.
      Sci. Part A: Polym. Chem., 1999, 37, 1365–1381.

[52]  Bauer, K.N., Tee, H.T., Lieberwirth, I., & Wurm, F.R. In-chain poly(phosphonate)s via acyclic
      diene metathesis polycondensation, Macromol., 2016, 49, 3761–3768.

[53]  Steinbach, T., Ritz, S., & Wurm, F.R. Water-soluble poly-(phosphonate)s via living
      ring-opening polymerization, ACS. Macro. Lett., 2014, 3, 244–248.

[54]  Stoianova, D.S., & Hanson, P.R. A ring-closing metathesis strategy to phosphonosugars, Org.
      Lett., 2001, 3, 3285–3288.

[55]  Roy, R. Glycomimetics: Modern synthetic methodologies, Am. Chem. Soc., 2005, 896, 206.

[56]  Hanessian, S., Galéotti, N., Rosen, P., Oliva, G., & Babu, S. Synthesis of carbohydrate
      phostones as potential glycomimetics, Bioorg. Med. Chem. Lett., 1994, 4, 2763–2768.

[57]  Allcock, H.R. Chemistry and applications of polyphosphazenes, Wiley, Hoboken, USA, 2003.

[58]  Breza, M. On bonding in cyclic triphosphazenes, J. Mol. Struct. Theochem., 2000, 505,
      169–177.

[59]  Enlow, M. Ab-initio studies of cyclic phosphazine systems (NPX$_2$)$_n$: A study of the structure
      and bonding in such systems and a search for model systems for the polymer, Polyhedron,
      2003, 22, 473–482.

[60]  Allcock, H.R., Reeves, S.D., de Denusand, C.R., & Crane, C.K. Influence of reaction
      parameters on the living cationic polymerization of phosphoranimines to polyphosphazenes,
      Macromol., 2001, 34, 748–754.

[61]  Allcock, H.R., Nelson, J.M., Reeves, S.D., Honeyman, C.H., & Manners, I. Ambient temperature
      direct synthesis of poly(organophosphazenes) via the living cationic polymerization of
      organo-substituted phosphoranimines, Macromol., 1997, 30, 50–56.

[62]  Allcock, H.R., Reeves, S.D., Nelson, J.M., & Manners, I. Synthesis and characterization of
      phosphazene di- and tri-block copolymers via the controlled cationic, ambient temperature
      polymerization of phosphoranimines, Macromol., 2000, 33, 3999–4007.

[63] Sun, Q., Sun, J., Gong, S.S., & Wang, X.C. One-pot synthesis of symmetrical dinucleoside polyphosphates and analogs via 4,5-dicyanoimidazole-promoted tandem P–O coupling reactions, Tetrahedron Lett., 2014, 55, 5785–5788.

[64] Wang, Q., Ci, Z., Zhu, G., Xin, S., Zeng, W., Que, M., & Wang, Y. Multicolor bright $Ln^{3+}$ (Ln = Eu, Dy, Sm) activated tungstate phosphor for multifunctional applications, Opt. Mater. Express, 2014, 4, 142–154.

[65] Ehrenberg, H., Bramnik, N.N., Sensyshyn, A., & Fuess, H. Crystal and magnetic structures of electrochemically delithiated $Li_{1-x}$ $CoPO_4$ phases, Solid. State. Sci., 2009, 11, 18–23.

[66] Hatert, F., Fransolet, A.-M., & Maresh, W.V. The stability of primary alluaudites in granitic pegmatites: An experimental investigation of the $Na_2(Mn_{1-x}Fe^{2+}x)_2Fe^{3+}(PO_4)_3$ solid solution: Contribution to mineralogy and petrology, Mineral. Petrol., 2006, 152, 399–419.

[67] Ternane, R., Ferid, M., Panczer, G., Trabelsi-Ayadi, G., & Boulon, G. Site-selective spectroscopy of $Eu^{3+}$-doped orthorhombic lanthanum and monoclinic yttrium polyphosphates, Opt. Mater., 2005(27), 1832–1838.

[68] Teikemeier, G., & Goldberg, D.J. Skin resurfacing with the erbium: YAG laser, Dermatol Surg., 1997, 23, 685–687.

[69] Ferhi, M., & Horchani-Naifer, K. Spectroscopic properties of $Eu^{3+}$-doped $KLa(PO_3)_4$ and LiLa $(PO_3)_4$ powders, Opt. Mater., 2011, 34, 12–18.

[70] Chékir-Mzali, J., Horchani-Naifer, K., & Férid, M. Structural study and spectroscopic properties of $NH_4Er(PO_3)_4$, J. Alloys. Compd., 2014, 612, 372–379.

[71] Chemingui, S., Ferhi, M., Horchani-Naifer, K., & Férid, M. Synthesis, characterization and optical properties of $NH_4Dy(P\ O_3)_4$, J. Solid State Chem., 2014, 217, 99–104.

[72] Lukowiak, A., Wiglusz, R.J., Chiappini, A., Armellini, C., Battisha, I.K., Righini, G.C., & Ferrari, M. Structural and spectroscopic properties of $Eu^{3+}$-activated nanocrystalline tetraphosphates loaded in silica–Hafnia thin film, J. Non-Crystal. Solids, 2014, 401, 32–35.

[73] Chemingui, S., Ferhi, M., Horchani-Naifer, M., & Férid, M. Synthesis and luminescence characteristics of $Dy^{3+}$ doped $KLa(PO_3)_4$, J. Lumin, 2015, 166, 82–87.

[74] Chékir-Mzali, J., Horchani-Naifer, K., & Férid, M. Investigations of vibrational, structural and optical properties of erbium polyphosphate micro-powders, Optik, 2016, 127, 6340–6350.

[75] Ferhi, M., Ben Hassen, N., Bouzidi, C., Horchani-Naifer, K., & Ferid, M. Near-infrared luminescence properties of $Yb^{3+}$ doped $LiLa(PO_3)_4$ powders, J. Lumin, 2016, 170, 174–179.

[76] Abdelhedi, M., Horchani-Naifer, K., Dammak, M., & Ferid, M. Structural and spectroscopic properties of pure and doped $LiCe(PO_3)_4$, Mater. Res. Bull., 2015, 70, 303–308.

[77] Ben Hassen, N., Ferhi, M., Horchani-Naifer, K., & Férid, M. Structure determination and optical properties of $CsSm(PO_3)_4$, Mater. Res. Bull., 2015, 63, 99–104.

[78] Qin, Z., Li, D., Zhang, W., & Yang, R.. Surface modification of ammonium polyphosphate with vinyl trimethoxysilane: Preparation, characterization, and its flame retardancy in polypropylene, Polym. Degrad. Stab., 2015, 119, 139–150.

[79] Szymusiak, M., Donovan, A.J., Smith, S.A., Ransom, R., Shen, H., Kalkowski, J., Morrissey, J.H., & Liu, Y. Colloidal confinement of polyphosphate on gold nanoparticles robustly activates the contact pathway of blood coagulation, Bioconjugate Chem., 2016, 27, 102–109.

[80] Choi, S.H., Collins, J.N.R., Smith, S.A., Davis-Harrison, R.L., Rienstra, C.M., & Morrissey, J.H. Polyphosphates: Facile manipulation of polyphosphate for investigating and modulating its biological activities, Biochemistry, 2010, 49, 9935–9941.

[81] Kim, H.H., Lee, H., Kim, H.E., Seo, J., Hong, S.W., Lee, J.Y., & Lee, C. Polyphosphate-enhanced production of reactive oxidants by nanoparticulate zero-valent iron and ferrous ion in the presence of oxygen: Yield and nature of oxidants, Water Res., 2015, 86, 66–73.

[82] Iwasaki, Y., & Akiyoshi, K. Synthesis and characterization of amphiphilic polyphosphates with hydrophobic graft chains and cholesteryl groups as nanocarriers, Biomacromolecules, 2006, 7, 1433–1438.

[83] Gao, W., Wang, S., Ma, H., Wang, Y., & Meng, F. Combined situ polymerization and thermal cross-linking technique for the preparation of ammonium polyphosphate microcapsules a with composite shell, J. Phys. Chem. C, 2015, 119, 28999–29005.

[84] Liu, J., Huang, W., Pang, Y., Zhu, X., Zhou, Y., & Yan, D. Hyperbranched polyphosphates for drug delivery application: Design, synthesis and in vitro evaluation, Biomacromolecules, 2010, 11, 1564–1570.

[85] Donovan, A.J., Kalkowski, J., Smith, S.A., Morrissey, J.H., & Liu, Y. Size-controlled synthesis of granular polyphosphate nanoparticles at physiologic salt concentrations for blood clotting, Biomacromol., 2014, 15, 3976–3984.

[86] Alexandridis, P.M.T. Block copolymer-directed metal nanoparticle morphogenesis and organization, Eur. Polym. J., 2011, 47, 569–583.

[87] Galezowska, J.E.G. Phosphonates, their complexes and bio-applications: A spectrum of surprising, Coord. Chem. Rev., 2012, 256, 105–124.

[88] Monteil, M., Moustaoui, H., Picardi, G., Aouidat, F., Djaker, N., Chapelle, M.L.L., Lecouvey, M., & Spadavecchia, J. Polyphosphonate ligands: From synthesis to design of hybrid PEGylated nanoparticles toward phototherapy studies, J. Colloid Interface Sci., 2018, 513, 205–213.

[89] Wang, Y.C., Yuan, Y.Y., Du, J.Z., Yang, X.Z., & Wang, Y.Z. Recent progress in polyphosphoesters: From controlled synthesis to biomedical applications, J. Macromol. Biosci., 2009, 9(12), 1154–1164.

[90] Reese, C.B. Oligo- and poly-nucleotides: 50 years of chemical synthesis, Org. Biomol. Chem., 2005, 3, 3851–3868.

[91] Richards, M., Dahiyat, B.I., Arm, D.M., Brown, P.R., & Leong, K.W. Evaluation of polyphosphates and polyphosphonates as degradable biomaterials, J. Biomed. Mater. Res., 1991, 25, 1151–1167.

[92] Steinbach, T., Alexandrino, E.M., & Wurm, F.R. Unsaturated poly(phosphoester)s via ring-opening metathesis polymerization, Polym. Chem., 2013, 4(13), 3800–3806.

[93] Marsico, F., Turshatov, A., Weber, K., & Wurm, F.R. A metathesis route for BODIPY labeled polyolefins, Org. Lett., 2013, 15, 3844–3847.

[94] Iwasaki, Y., & Yamaguchi, E. Synthesis of well-defined thermo responsive polyphosphoester macroinitiators using organocatalysts, Macromolecules, 2010, 43, 2664–2666.

[95] Marsico, F., Wagner, M., Landfester, K., & Wurm, F.R. Unsaturated polyphosphoesters via acyclic diene metathesis polymerization, Macromolecules, 2012, 45, 8511–8518.

[96] Zhang, S., Li, A., Zou, J., Lin, L., & Wooley, K. Facile synthesis of clickable, water-soluble and degradable polyphosphoesters, ACS. Macro. Lett., 2012, 1, 328–333.

[97] Zhang, S., Zou, J., Zhang, F., Elsabahy, M., Felder, S., Zhu, J., Pochan, D., & Wooley, K. Rapid and versatile construction of diverse and functional nanostructures derived from a polyphosphoester-based biomimetic block copolymer system, J. Am. Chem. Soc., 2012, 134, 18467–18474.

[98] Zhang, S., Wang, H., Shen, Y., Zhang, F., Seetho, K., Zou, J., Taylor, J.S.A., Dove, A.P., & Wooley, K.L. A simple and efficient synthesis of an acid-labile polyphosphoramidate by organo base-catalyzed ring-opening polymerization and transformation to polyphosphoester ionomers by acid treatment, Macromolecules, 2013, 46(13), 5141–5149.

[99] Mauldin, T.C., Zammarano, M., Gilman, J.W., Shields, J.R., & Boday, D.J. Synthesis and characterization of isosorbide-based polyphosphonates as biobased flame-retardants, Polym. Chem., 2014, 5, 5139–5146.

[100] Steinbach, T., Ritz, S., Frederik, R., & Wurm, R. Water-soluble poly(phosphonate)s via living ring-opening polymerization, ACS. Macro. Lett., 2014, 3, 244–248.

[101] Allcock, H.R. Recent advances in phosphazene (phosphonitrilic) chemistry, Chem. Rev., 1972, 72, 315–356.

[102] Tian, Z., Hess, A., Fellin, C.R., Nulwala, H., & Allcock, H.R. Phosphazene high polymers and models with cyclic aliphatic side groups: New structure–Property relationships, Macromolecules, 2015, 48, 4301–4311.

[103] Liu, X., Breon, J.P., Chen, C., & Allcock, H.R. Substituent exchange reactions with high polymeric organophosphazenes, Macromolecules, 2012, 45, 9100–9109.

[104] Carriedo, G.A., Alonso, F.L.G., Gomez-Elipe, P., Fidalgo, J.I., Alvarez, J.L.G., & Presa Soto, A. A simplified and convenient laboratory-scale preparation of $^{14}$N or $^{15}$N high molecular weight poly(dichlorophosphazene) directly from PCl$_5$, Chem. A Europ. J., 2003, 9, 3833–3836.

[105] Blackstone, S., Pfirrmann, H., Helten, A., Staubitz, A., Presa Soto, G., Whittell, R., & Manners, I. A cooperative role for the counter anion in the PCl$_5$-initiated living, cationic chain growth polycondensation of the phosphoranimine Cl$_3$P—NSiMe$_3$, J. Am. Chem. Soc., 2012, 134, 15293–15296.

[106] Wilfert, S., Henke, H., Schoefberger, W., Bruggemann, O., & Teasdale, I. Chain-end-functionalized polyphosphazenes via a one-pot phosphine-mediated living polymerization, Macromol. Rapid Commun., 2014, 35, 1135–1141.

[107] Steinke, J.H.G., Greenland, B.W., Johns, S., Parker, M.P., Atkinson, R.C.J., Cade, I.A., Golding, P., & Trussell, S.J. Robust and operationally simple synthesis of poly(bis(2,2,2-trifluoroethoxy) phosphazene) with controlled molecular weight, low PDI and high conversion, ACS. Macro. Lett., 2014, 3, 548–551.

[108] Soto, A.P., & Manners, I. Poly(ferrocenylsilane-b-polyphosphazene) (PFS-b-PP): A new class of organometallic–inorganic block copolymers, Macromol, 2009, 42, 40–42.

[109] Allcock, H.R., de Denus, C.R., Prange, R., & Laredo, W.R. Synthesis of nor bornenyl telechelic polyphosphazenes and ring-opening metathesis polymerization reactions, Macromolecules, 2001, 34, 2757–2765.

[110] Henke, H., Wilfert, S., Iturmendi, A., Bruggemann, O., & Teasdale, I. Branched polyphosphazenes with controlled dimensions, J. Polym. Sci., Part A: Polym. Chem., 2013, 51, 4467–4473.

[111] Liu, X., Tian, Z., Chen, C., & Allcock, H.R. Synthesis and characterization of brush-shaped hybrid inorganic/organic polymers based on polyphosphazenes, Macromolecules, 2012, 45, 1417–1426.

[112] Allcock, H. R. Polyphosphazene elastomers, gels and other soft materials, Soft Matter, 2012, 8, 7521–7532.

[113] Alexander, S. H. Method of treating asphalt, U. S. Patent, 1973, 3(751), 278.

[114] Giavarini, C., Mastrofini, D., & Scarsella, M. Macrostructure and rheological properties of chemically modified residues and bitumens, Energy Fuels, 2000, 14, 495–502.

[115] Orange, G., Dupuis, D., Martin, J. V., Farcas, F., Such, C., & Marcant, B. Chemical modification of bitumen through polyphosphoric acid: Properties-microstructure relationship, Proceedings of the 3$^{rd}$ Euraphalt and Eurobitume Congress, Vienna, Austria, 2004, paper 334, book 1, 733–745.

[116] Chebuliez, E., & Weniger, H. Phosphorylation sparles acid espolyphosphoriques, He IV. Chim. Acta, 1946, 29, 2006–2017.

[117] Koebner, A., & Robinson, R. Experiments on the synthesis of substances related to the sterols. Part XXII. Synthesis of X-norequilen in methyl ether, J. Chem. Soc., 1938, 1994–1997.

[118] Chebuliez, E. Organic derivatives of phosphoric acid, Organic Phosphorus Compounds, Kosolap Off, G.M., Maier, L., Eds, Wiley-Interscience, NewYork, 1973.

[119] Rowlands, D.A. Polyphosphoric acid (PPA), In Synthetic Reagents, Pizey, J.S., Ed.;, Ellis Horwood Publishers, Chichester, U.K, 1985, 6.

[120] Averbuch-Pouchot, M.T., & Durif, A. Topics in Phosphate Chemistry, World Scientific, Hackensack, NJ, 1996.

[121] Corbridge, D.E.C. Phosphorus: An outline of its chemistry. Biochem. Technol, 5th ed., Elsevier, New York, 1995.

[122] Jameson, R.F. The composition of the strong phosphoric acid, J. Chem. Soc., 1959, 752–759.

[123] Allcock, H.R., Kugel, R.L., & Stroh, E.G. The structure and properties of poly (difluorophosphazenes), Inorg. Chem., 1972, 11, 1120–1123.

[124] Allcock, H.R., Connolly, M.S., Sisko, J.T., & Al-Shali, S. Effects of organic side group structures on the properties of poly(organophosphazenes), Macromolecules, 1988, 21, 323–334.

[125] Becker, Pierre. Phosphates and phosphoric acid, Marcel Dekker, New York, ISBN 0824717120, 1988.

[126] Norman, N., & Earnshaw, A. Chemistry of the Elements, 2nd, Butterworth -Heinemann, 1997, 520–522, ISBN0-08-037941-9.

[127] Popp, F.D., & McEwen, W.E. Polyphosphoric acid as a reagent in organic chemistry, Chem. Rev., 1959, 58, 321–401.

[128] Salah, A.M., Sabot, R., Refait, P., Liascukiene, I., Méthivier, C., Landoulsi, J., Dhouibi, L., & Jeannin, M. Passivation behaviour of stainless steel (UNS N-08028) in industrial or simplified phosphoric acid solutions at different temperatures, Corros. Sci., 2015, 99, 320–332.

[129] Sleymi, S., Kahlaoui, M., Dkhili, S., Besbes-Hentati, S., & Abid, S. Synthesis, crystal structure, characterization and electrochemical properties of a new cyclohexaphosphate: $Li_2Na_2CoP_6O_{18} \cdot 12H_2O$, J. Mol. Struct., 2017, 1127, 175–182.

[130] Yang, JC., Liao, W., Deng, SB., Cao, ZJ., & Wang, YZ. Flame retardation of cellulose-rich fabrics via a simplified layer-by-layer assembly, Carbohydr. Polym., 2016, 151, 434–440.

[131] Kim, N.K., Lin, R.J.T., & Bhattacharyya, D. Effects of wool fibres, ammonium polyphosphate and polymer viscosity on the flammability and mechanical performance of PP/wool composites, Polym. Degrad. Stab., 2015, 119, 167–177.

[132] Qin, Z., Li, D., & Yang, R. Study on inorganic modified ammonium polyphosphate with precipitation method and its effect in flame retardant polypropylene, Polym. Degrad. Stab., 2016, 126, 117–124.

[133] Leistner, M., Haile, M., Rohmer, S., Abu-Odeh, A., & Grunlan, J.C. Water-soluble polyelectrolyte complex nanocoating for flame retardant nylon-cotton fabric, Polym. Degrad. Stab., 2015, 122, 1–7.

[134] Yang, J.C., Cao, Z.J., Wang, Y.Z., & Schiraldi, D.A. Ammonium polyphosphate-based nanocoating for melamine foam towards high flame retardancy and anti-shrinkage in fire, Polymer, 2015, 66, 86–93.

[135] Feng, A., Liang, M., Jiang, J., Huang, J., & Liu, H. Synergistic effect of a novel triazine charring agent and ammonium polyphosphate on the flame retardant properties of halogen-free flame retardant polypropylene composites, Thermochim. Acta, 2016, 627–629, 83–90.

[136] Xie, H., Lai, X., Zhou, R., Li, H., Zhang, Y., Zeng, X., & Guo, J. Effect and mechanism of N-alkoxy hindered amine on the flame retardancy, UV aging resistance and thermal degradation of intumescent flame retardant polypropylene, Polym. Degrad. Stab., 2015, 118, 167–177.

[137] Mariappan, T., & Wilkie, C.A. Flame retardant epoxy resin for electrical and electronic applications, Fire Mater, 2014, 38, 588–598.

[138] Sharmila, T.K.B., Nair, A.B., Abraham, B.T., Beegum, P.M.S., & Thachil, E.T. Microwave exfoliated reduced graphene oxide epoxy nanocomposites for high performance applications, Polymer, 2014, 55, 3614–3627.

[139] Szyndler, M.W., Timmons, J.C., Yang, Z.H., Lesser, A.J., & Emrick, T. Multifunctional deoxybenzoin-based epoxies: Synthesis, mechanical properties, and thermal evaluation, Polymer, 2014, 55, 4441–4446.

[140] Zhao, W., Liu, J., Peng, H., Liao, J., & Wang, X. Synthesis of a novel PEPA-substituted polyphosphoramide with high char residues and its performance as an intumescent flame retardant for epoxy resins, Polym. Degrad. Stab., 2015, 118, 120–129.

[141] Tan, Y., Shao, Z.B., Yu, L.X., Xu, Y.J., Rao, H., Li, C., & Wang, Y.Z. Polyethyleneimine modified ammonium polyphosphate toward polyamine-hardener for epoxy resin: Thermal stability, flame retardance and smoke suppression, Polym. Degrad. Stab., 2016, 131, 62–70.

[142] Verdolotti, L., Oliviero, M., Lavorgna, M., Iannace, S., Camino, G., Vollaro, P., & Frache, A. On revealing the effect of alkaline lignin and ammonium polyphosphate additives on fire retardant properties of sustainable zein-based composites, Polym. Degrad. Stab., 2016, 134, 115–125.

[143] Zhang, W., He, X., Song, T., Jiao, Q., & Yang, R. Comparison of intumescence mechanism and blowing-out effect in flame-retarded epoxy resins, Polym. Degrad. Stab., 2015, 112, 43–53.

[144] Chivas-Joly, A., Longuet, C., Motzkus, C., & Lopez-Cuesta, J.M. Influence of the composition of PMMA nanocomposites on gaseous effluents emitted during combustion, Polym. Degrad. Stab., 2015, 113, 197–207.

[145] Xie, B., Wang, Y.Z., Yang, B., & Liu, Y. A novel intumescent flame-retardant polyethylene system, Macromol. Mater. Eng., 2006, 291, 247–253.

[146] Sheng, H., Zhang, Y., Wang, B., Yu, B., Shi, Y., Song, L., Kundu, C.K., Tao, Y., Jie, G., Feng, H., & Hu, Y. Effect of electron beam irradiation and microencapsulation on the flame retardancy of ethylene-vinyl acetate copolymer materials during hot water ageing test, Radiat. Phys. Chem., 2017, 133, 1–98.

[147] Lesaffre, N., Bellayer, S., Fontaine, G., Jimenez, M., & Bourbigot, S. Revealing the impact of ageing on a flame retarded PLA, Polym. Degrad. Stab., 2016, 127, 88–97.

[148] Bcoz, K., Domonkos, M., Igricz, T., Kmetty, A., Bárány, T., & Marosi, G. Flame retarded self-reinforced poly(lactic acid) composites of outstanding impact resistance, Composites: Part A, 2015, 70, 27–34.

[149] Ramasahayam, S.K., Clark, A.L., Hicks, Z., & Viswanathan, T. Spent coffee grounds derived P, N co-doped C as electrocatalyst for supercapacitor applications, Electrochim. Acta, 2015, 168, 414–422.

[150] Muller, W.E.G., Tolba, E., Schröder, H.C., Wang, S., Glaßer, G., Muñoz-Espí, R., Link, T., & Wang, X. A new polyphosphate calcium material with morphogenetic activity, Mater. Lett., 2015, 148, 163–166.

[151] Müller, G., Neufurth, M., Tolba, E., Wang, W., Geurtsen, W., Feng, Q.L., Schroder, HC., & Wang, XH. Biomimetic approach to ameliorate dental hypersensitivity by amorphous polyphosphate microparticles, Dental Mater., 2016, 32, 775–783.

[152] Müller, W.E.G., Tolba, E., Schröder, H.C., Diehl-Seifert, B., & Wang, X. Retinol encapsulated into amorphous $Ca^{2+}$ polyphosphate nanospheres acts synergistically in MC3T3-E1 cells, Eur. J. Pharm.Biopharm., 2015, 93, 214–223.

[153] doAmaral, J.G., Delbem, A.C.B., Pessan, J.P., Manarelli, M.M., & Barbour, M.E. Effects of polyphosphates and fluoride on hydroxyapatite dissolution: A pH-stat investigation, Arch. Oral Bio., 2016, 63, 40–46.

[154] Ozeki, N., Hase, N., Yamaguchi, H., Hiyama, T., Kawai, R., Kondo, A., Nakataand, K., & Mogi, M. Polyphosphate induces matrix metalloproteinase-3-mediated proliferation of odontoblast-like cells derived from induced pluripotent stem cells, Exp. cell res., 2015, 333, 303–315.

[155] Sanderson, J., Dartt, D.A., Trinkaus-Randall, V., Pintor, J., Civan, M.M., Delamere, N.A., Fletcher, E.L., Salt, T.E., Grosche, A., & Mitchell, C.H. Purines in the eye: Recent evidence for the physiological and pathological role of purines in the RPE, retinal neurons, astrocytes, muller cells, lens, trabecular meshwork, cornea and lacrimal gland, Exp. cell res., 2014, 127, 270–287.

[156] Rodrigues, S., Cordeiro, C., Seijo, B., Remu˜nán-López, C., & Grenha, A. Hybrid nanosystems based on natural polymers as protein carriers for respiratory delivery: Stability and toxicological evaluation, Carbohydr.Polym., 2015, 123, 369–380.

[157] Walke, S., Srivastava, G., Nikalje, M., Doshi, J., Kumar, R., Ravetkar, S., & Doshi, P. Fabrication of chitosan microspheres using vanillin/TPP dual crosslinkers for protein antigens encapsulation, Carbohydr. Polym., 2015, 128, 188–198.

[158] Xue, M., Hu, S., Lu, Y., Zhang, Y., Jiang, X., An, S., Guo, Y., Zhou, X., Hou, H., & Jiang, C. Development of chitosan nanoparticles as drug delivery system for a prototype capsid inhibitor, Int. J. Pharm., 2015, 495, 771–782.

[159] Prataand, A.S., & Grosso, C.R.F. Production of microparticles with gelatin and chitosan, Carbohydr. Polym., 2015, 116, 292–299.

[160] Yadollahi, M., Farhoudian, S., & Namazi, H. One-pot synthesis of antibacterial chitosan/ silver bio-nanocomposite hydrogel beads as drug delivery systems, Int. J. Biol. Macromol., 2015, 79, 37–43.

[161] Zhang, A., Sun, B., Li, X., Yu, Y., Tian, Y., Xu, X., & Jin, Z. Synthesis of pH- and ionic strength- responsive microgels and their interactions with lysozyme, Int. J. Biol. Macromol., 2015, 79, 392–397.

[162] Zhang, A., Wei, B., Hu, X., Jin, Z., Xu, X., & Tian, Y. Preparation and characterization of carboxymethyl starch microgel with different crosslinking densities, Carbohydr. Polym., 2015, 124, 245–253.

[163] Wintgens, V., Lorthioir, C., Dubot, P., Sébille, B., & Amiel, C. Cyclodextrin/dextran based hydrogels prepared by cross-linking with sodium trimetaphosphate, Carbohydr. Polym., 2015, 132, 80–88.

[164] Liu, J., Pang, Y., Huang, W., Zhu, Z., Zhu, X., Zhou, Y., & Yan, D. Redox-responsive polyphosphate nanosized assemblies: A smart drug delivery platform for cancer therapy, Biomacromolecules, 2011, 12, 2407–2415.

[165] Li, Y., Wang, J., Du, W., & Wang, S. Transplantation of copper-doped calcium polyphosphate scaffolds combined with copper (II) preconditioned bone marrow mesenchymal stem cells for bone defect repair, J. Biomat. Appl., 2018, 32, 738–753.

[166] Welles, L., Abbas, B., Sorokin, D.Y., Hooijmans, C.M., van Loosdrecht, M.C.M., & Brjanovic, C. Metabolic response of " Accumulibacter Phosphatis" clade II C to changes in influent P/C ratio, Front Microbiol., 2016, 7(2121), 1–16.

[167] Lanham, A.B., Oehmen, A., Saunders, A.M., Carvalho, G., Nielsen, P.H., & Reis, M.A.M. Metabolic modelling of full-scale enhanced biological phosphorus removal sludge, Water Res., 2014, 66, 283–295.

[168] Motlagh, M., Bhattacharjee, A.S., & Goel, R. Microbiological study of bacteriophage induction in the presence of chemical stress factors in enhanced biological phosphorus removal (EBPR), Water Res., 2015, 81, 1–14.

[169] Carvalheira, M., Oehmen, A., Carvalho, G., Eusebio, M., & Reis, M.A.M. The impact of aeration on the competition between polyphosphate accumulating organisms and glycogen accumulating organisms, Water Res., 2014, 66, 296–307.

[170] Ge, A., Batstone, D.J., & Keller, J. Biological phosphorus removal from abattoir wastewater at very short sludge ages mediated by novel PAO clade Comamonadaceae, Water Res., 2015, 69, 173–182.

[171] Taya, A., Guerrero, J., Suárez-Ojeda, M.E., Guisasola, A., & Baeza, J.A. Assessment of crude glycerol for enhanced biological phosphorus removal: Stability and role of long chain fatty acids, Chemosphere, 2015, 141, 50–56.

[172] Fernández-Fernández, M., Gómez-Rey, M.X., & González-Prieto, S.J. Effects of fire and three fire-fighting chemicals on main soil properties, plant nutrient content and vegetation growth and cover after 10 years, Sci. Total Environ., 2015, 515–516, 92–100.

[173] Paker, A., & Matak, K.E. Impact of sarcoplasmic protein on texture and color of silver carp and Alaska pollock protein gels, LWT – Food Sci. Technol., 2015, 63, 985–991.

[174] Cadavez, V., Gonzales-Barron, U., Pires, P., Fernandes, E., Pereira, A.P., Gomes, A., Araújo, J.P., Lopes-da-Silva, F., Rodrigues, P., Fernandes, C., Saavedra, M.J., Butler, F., & Dias, T. An assessment of the processing and physicochemical factors contributing to the microbial contamination of salpicao, a naturally-fermented Portuguese sausage, LWT – Food Sci. Technol., 2016, 72, 107–116.

[175] García-García, E., & Totosaus, A. Low-fat sodium-reduced sausages: Effect of the interaction between locust bean gum, potato starch and κ-carrageenan by a mixture design approach, Meat Sci., 2008, 78, 406–413.

[176] Cofrades, S., López-López, I., Solas, M.T., Bravo, L., & Jiménez-Colmenero, F. Influence of different types and proportions of added edible seaweeds on characteristics of low-salt gel/emulsion meat systems, Meat Sci., 2008, 79, 767–776.

[177] Marchetti, L., Argel, N., Andrés, S.C., & Califano, A.N. Sodium-reduced lean sausages with fish oil optimized by a mixture design approach, Meat Sci., 2015, 104, 67–77.

[178] Sathiyabama, M., & Einstein Charles, R. Fungal cell polymer-basedbased nanoparticle the s in protection of tomato plants from wilt disease caused by Fusarium oxysporum f. sp. lycopersici, Carbohydr. Polym., 2015, 133, 400–407.

[179] Gray, M.J., & Jakob, U. Oxidative stress protection by polyphosphate – New roles for an old player, Curr. Opin. Microbiol., 2015, 24, 1–6.

[180] Racki, L.R., Tocheva, E.I., Dieterle, M.G., Sullivan, M.C., Jensen, G.J., & Newman, D.K. Polyphosphate granule biogenesis is temporally and functionally tied to cell cycle exit during starvation in Pseudomonas aeruginosa, Proc. Natl Acad. Sci. U S A, 2017, 114, 2440–2449.

[181] Schulz, H.N., & Schulz, H.D. Large sulfur bacteria and the formation of phosphorite, Science, 2005, 307, 416–418.

[182] Crosby, C.H., & Bailey, J.V. The role of microbes in the formation of modern and ancient phosphatic mineral depo sits, Front. Microbiol., 2012, 241, 3–9.

[183] Breilanda, A.A., Flooda, B.E., Nikra, d J., Bakaric, h J., Husman, M., Rhee, T.H., Jones, R.S., & Bailey, J.V. Polyphosphate-accumulating bacteria: Potential contributors to mineral dissolution in the oral cavity, Appl. Environ. Microbiol., 2018, 84, In Press.

[184] Omelon, S.J., & Grynpas, M.D. Relationships between polyphosphate chemistry, biochemistry and apatite biomineralization, Chem. Rev., 2008, 108, 4694–4715.

[185] Kaniappan, K., Murugavel, S.C., & Thangadurai, T.D. Synthesis and properties of few polyphosphonate derivatives containing photosensitive unsaturated keto group in the main chain, Macromol. Res., 2013, 21, 1045–1053.

# 4 Sulfur-based inorganic polymers: polythiazyl and polythiol

**Abstract:** Polythiazyl $(SN)_x$ can be synthesized by the following methods, that is, electrochemical oxidation, click reaction thiol-ene reaction, and so on. Polythiazyl polymers are analyzed by differential scanning calorimetry and $^1H$ NMR spectroscopic methods and are used as healing agents. Polythiols are reactive compounds with several mercaptans having various applications. This review gives a brief knowledge of polythiazyl, parathiocyanogen, polythiol, polysulfur, polysulfide, and polysulfone ranging from their synthesis and uses in biomedical field.

**Keywords:** polythiazyl, polythiols, thermal stability

## 4.1 Introduction

Polythiazyl is actually an electrically conductive polymer and also possesses a very good metal luster. It is often golden or maybe bronze in color and behaves as a superconductor below 0.28 K. It was the very first discovered conductive and inorganic polymer [1, 2]. It is air stable and insoluble in all of the solvents [3]. It possesses N and S atoms within the adjacent chain of the polymeric framework [4] (Figure 4.1). Polythiazyl is prepared by the polymerization of the disulfur dinitride $(S_2N_2)$. Polythiazyl is utilized in light-emitting diodes (LEDs), solar cells, and transistors due to its very good electrical conductivity [5].

Parathiocyanogen or polythiocyanogen is an inorganic polymer that is formed by spontaneous polymerization of thiocyanogen $(SCN)_2$. Parathiocyanogen is known to have an empirical formula $(SCN)_x$. It is a brick-red or orange amorphous powder and is insoluble in all common solvents. Cataldo investigated new data from $^{13}C$ NMR spectrometry as well as from FT-IR (Fourier-transform infrared), UV-VIS, X-ray diffraction pattern, and thermal stability (thermogravimetry-differential thermal analysis), which confirmed that parathiocyanogen or polythiocyanogen $(SCN)_x$ has a linear structure analogous to that assigned to polythiazyl or polysulfur nitride $(SN)_x$. Based also on the chemical reaction of $(SCN)_x$ with NaCN and $Na_2S$, it is shown that the other alternative structures that are taken into account (sym-triazine based and polyazomethine with sulfur bridges) are not likely the correct ones for parathiocyanogen [6]. The evidence that parathiocyanogen has all its monomeric units connected in a head-to-tail fashion comes from a study of the $^{13}C$ NMR spectrum, where the carbon atom signal appears as a singlet at 186 ppm. Cataldo found that this signal is at considerably lower magnetic fields. The chemical shift found at 186 ppm [in dimethyl formamide (DMF)] for parathiocyanogen is consistent with the linear conjugated structure and rules out the cyclic triazine as

https://doi.org/10.1515/9781501514609-005

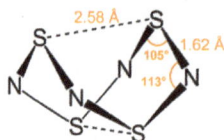

**Figure 4.1:** Structure of polythiazyl.

well as the polyazomethine (polynitrile) structures. The latter structures may be reasonably proposed for this material since the maximum chemical shift for the carbon atom of a polyazomethine chain is expected at 165 ppm [7]. Nitriles, isonitriles, thiocyanides, and isothiocyanides have their chemical shifts at fields considerably higher than that observed for $(SCN)_x$.

Polythiols were used as adhesives and chain transfer agents. They were also used as inhibitors against corrosion and in cosmetics [8]. The thiolate anions and thiyl radicals react through "click" reactions [9, 10]. The reactivity of thiols was affected by the $pK_a$ value of thiols [11, 12]. Polysulfides are the manufactured rubbers, which are principally utilized as sealants for cars, for construction, for marine uses, and as flexibilizing hardeners for epoxy cement. Their sulfur linkage provides good strength along with free rotation, resulting in a flexible and strong polymer. Atmospheric moisture can easily trigger the curing of polysulfides. It can also cure with the help of metals at room temperature and achieve the most extreme quality in 3–7 days.

Polysulfones are a group of thermoplastic polymers. These polymers are known for their durability and toughness at high temperatures. They contain the subunit aryl-$SO_2$-aryl, the characterizing highlight of which is the sulfone groups. The processing and high cost of ingredients, polysulfones are used in specialty applications and are often a superior replacement for polycarbonates. The expression "Polysulfone" is typically utilized for polyarylethersulfones (PAES), since just aromatic polysulfones are utilized in a specialized setting. Moreover, since ether groups are constantly present in the modernly utilized polysulfones, PAES are additionally required to as polyether sulfones, poly(arylene sulfone)s, or basically polysulfone (PSU). Poly(phenylene sulfone) can be prepared by Friedel–Crafts reaction from benzene sulfonyl chloride (Figure 4.2).

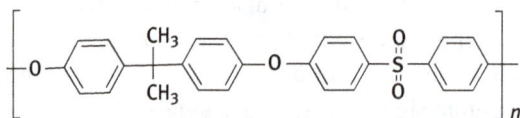

**Figure 4.2:** Structure of polysulfone.

## 4.2 Synthetic methods

### 4.2.1 Polythiazyl (SN)$_x$

#### 4.2.1.1 Epitaxial polymerization of (SN)$_x$

Orienting and ordering of polymer crystallization via epitaxy has a consistent and continuous rising number of applications, which extends from basic studies to higher approaches. Moreover, their crystallization in the form of prominent aniso-tropic folded chain lamellae reveals the orientation(s) induced by epitaxy. The caliber of polymer epitaxy in this approach is best declared by the "polymer deco-ration" technique.

Ishida et al. reported about the polymerization of epitaxially polymerized (SN)$_x$ [13]. Rickert et al. studied the preparation of disulfur dinitride epitaxially by using monovalent alkali halide single crystal. (SN)$_x$ was also prepared from the crystals that grow on NaCl, KBr, and KI [14].

#### 4.2.1.2 Solid-state polymerization

In the 1960s, solid-state polymerization was one of the well-known topics in the field of polymer science, which has been explored. By the use of high energy radiation, such as γ-rays, X-rays, electron beams, and sometimes α particle developed a solid-state polymerization. And also few polymerizations can be induced by visible light, UV radiation, thermally, and with the help of sensitizers. The term "solid-state poly-merization" covers versatile polymerization in the solid state, like crystalline, liquid crystalline, and glassy states of pure monomers, mixtures, or complexes. Chiang et al. reported the synthesis of tetrasulfur tetranitride ($S_4N_4$) and conducting metallic com-pounds like (SNBr$_{0.4}$)$_x$ and [SN(ICl)$_{0.4}$]$_x$ [15]. The (SN)$_x$ crystals were prepared by using $S_2N_2$ and solid-state polymerization method for 80 h at 10 °C [16, 17]. Sonnenschein and Seubert studied the synthesis of sulfobetaine exchangers by using quaternary amines and sulfonic acids [18].

### 4.2.2 Polythiocyanogen

According to Cataldo, thiocyanogen was synthesized using the following two routes:

$$\text{Pb (SCN)} + \text{Br}_2 \rightarrow \text{PbBr}_2 + \text{NC} - \text{S} - \text{S} - \text{CN} \tag{4.1}$$

$$2\,\text{AgCN} + \text{S}_2\text{Cl}_2 \rightarrow 2\,\text{AgCl} + \text{NC} - \text{S} - \text{S} - \text{CN} \tag{4.2}$$

In the first reaction [eq. (4.1)], the thiocyanogen formed is converted into parathio-cyanogen by a sudden exothermic reaction when the solvent is removed by distilla-tion. Thiocyanogen obtained by the second reaction [eq. (4.2)] is isolated by vacuum distillation from a dichloromethane solution at room temperature. The FT-IR spectra of parathiocyanogen obtained by the different routes were identical. The spectra of parathiocyanogen and its band assignments in relation to the linear structure have been discussed previously by Cataldo and Fiordiponti [19]. The spectra of trithiocyanuric acid and its polymeric derivative were obtained by oxidizing the mercapto groups with iodine. These two model compounds have infrared spectra of typical triazine-based compounds and are completely different from those of parathiocyanogen. The peculiar structure of parathiocyanogen is confirmed by X-ray diffraction spectra. On the other hand, the parathiocyanogen isomer with the triazine structure shows a completely different diffraction pattern. This observation is consistent with the fact that parathiocyanogen does not have a triazine-based structure. It was also shown that polymerization of thiocyanogen NC-S-S-CN involves the reaction of nitrile groups and that the monomer is a free radical formed by homolysis of S–S bonds of thiocyanogen. Furthermore, the polymerization appears to be autocatalytic, with a polymerization rate that depends on the concentration of thiocyanogen.

Baryshnikov et al. [20] reported the preparation of the polythiocyanogen-like (aryl-SCN)$_n$ (polythiocyanato hydroquinone) by using 1,4-benzoquinone and $NH_4SCN$ in the presence of glacial acetic acid. Polythiocyanatohydroquinone was used for the synthesis of composite materials of metal and also for making Pt electrode. Jones investigated the thermodynamics and decomposition kinetics of nitric oxide and (SN)$_x$, by using MNDO (modified neglect of diatomic overlap) and Austin model 1 [21].

Jones studied the open structure of NO, that is, ONNO and NNO and compared with the structure of SN. Therefore, it shows that the decomposition of the metastable polymer of nitrous oxide monomer was formed at extreme pressures [22]. Ancelin et al. reported about the hydrolysis of (SN)$_x$ and observed bands at different peaks [23]. Ramakrishnan and Chien investigated that phenylene-dithio-bis(phenyldithiazyl) and phenylthiophenyldithiazyl were good insulators [24].

### 4.2.3 Polythiol

Polymers containing free thiol as a functional group were prepared through chainend modification. Polymers containing dithio-moiety at one end were prepared through RAFT polymerization [25]. Nicolay reported the preparation of polyethylene copolymers through RAFT polymerization [26]. Methacrylate monomer comprising alkyl having sulfur and ethyl having oxygen xanthate moiety and thiol as a protecting group is prepared through RAFT polymerization. Polythiols are prepared by the polymerization of methacrylate and then aminolysis of the protecting groups occurs. Polythiol is made functional by using methods like thiol–ene addition, Michael addition, and thiol–disulfide exchange.

Tri-octylphosphine oxide (TOPO)-based quantum dots (QDs) are used for the application of biomedical imaging. The main drawback of TOPO-based QDs is the deactivation of the protein [27].

Aokia et al. studied the preparation of polyenes containing allyl-terminated (DAL) and norbornene-terminated DNb. As the number of terminal ene units increases, the photosensitivity of DAL/SH6 and DNb/SH6 resins increases [28]. Pahimanolisa et al. studied the preparation of thioether xylans by using water [29, 30]. Tao et al. investigated the synthesis of Ag-doped nanowires of polystyrene through polymerization method [31]. Isotropic conductive adhesives possess good electrical properties. Yuan et al. reported about the preparation of microcapsules for epoxy resin, which contains a curing agent by using poly(melamine–formaldehyde) and polythiol [pentaerythritol tetrakis (3-mercaptopropionate), PETMP] through in situ polymerization method [32]. Muench et al. reported the synthesis of thin films of nanocomposite by using ligand and metal nanoparticles [33]. Correa et al. studied the synthesis of poly-3-mercapto-propyl methylsiloxane film through click reaction [34].

### 4.2.4 Polysulfide

Polysulfide rubbers (T) are synthesized by alkyl halides and sodium polysulfide. Small particles of polymer precipitates are washed, coagulated, and dried. Straight polysulfides are synthesized by copolymerizing ethylene dichloride and di(chloroethyl) formal with the sulfur linkage. Branched polysulfides are produced using di (chloroethyl) formal and 2% trichloropropane. Vulcanized polysulfides are utilized for their great blend of low-temperature adaptability, gas and water impermeability, and protection from ozone, light, and heat. Expanded polymers are cured by peroxide with cross-linking at terminal thiols. Liquid polysulfide polymers with thiol terminals are also arranged. These can be cured at room temperature using peroxides or other oxygen giving restoring specialists properties like the other polysulfide rubbers.

Polysulfides can be prepared by the partial oxidation of sulfides. The fast processes for $HS^{n-}$ generation can be given as follows:

$$HS^- \rightleftharpoons HS^{2-} \rightleftharpoons HS^{3-} \rightleftharpoons \ldots \rightleftharpoons HS^{9-} \rightarrow S_8 + HS^-$$

Sulfur/polysulfide mixtures may be formed at pH $\leq$ 8, usually on the second timescale, depending on the medium and relative reactant concentrations. Due to fast reactivity of disulfides, the colloidal sols have been proposed as responsible for the onset of the 412 nm band assigned to Yellow. The physical and chemical properties of aqueous polysulfides show most intense electronic absorptions at wavelengths $\leq$300 nm; much weaker bands with maximum wavelengths <400 nm for different $HS^{n-}$ species ($n = 2$–5). Polysulfide polymers having structures $HS-(-C_2H_4OCH_2OC_2H_4S-S-)_n-C_2H_4OCH_2OC_2H_4-SH$ are readily available curing and modifying epoxies. It imparts impact resistance and toughness, increased flexibility, and reduced shrinkage.

### 4.2.5 Polysulfone

Polyethersulfones are synthesized using sodium salt of an aromatic diphenol and bis(4-chlorophenyl)sulfone by polycondensation (Figure 4.3). The sodium salt of the diphenol is prepared in situ by responding with a stoichiometries measure of sodium hydroxide (NaOH). Water must be expelled with an azeotropic dissolvable (e.g., methylbenzene or chlorobenzene) (Figure 4.3). The polymerization is completed at 130–160 °C under inert conditions in a polar, aprotic dissolvable, for example, dimethyl sulfoxide, forming a polyether by the disposal of sodium chloride.

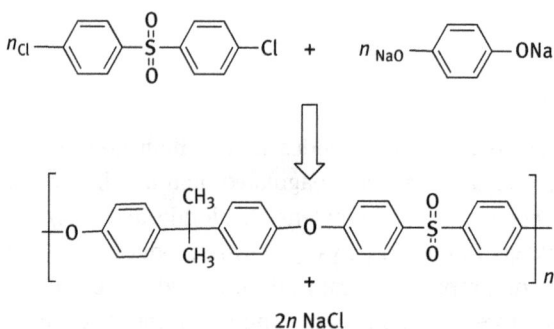

Figure 4.3: Synthesis of polysulfone.

Additionally, bis(4-fluorophenyl)sulfone can be utilized. It is more reactive than the dichloride, however, unreasonably costly for commercial uses. Through chain eliminators (e.g., chloromethane), the chain length can be directed at a range that a specialized dissolve preparing is conceivable. However, reactive end groups are still present in the product. To prevent further condensation in the melt, the end groups can be etherified with chloromethane. The diphenol is ordinarily bisphenol-An or 1,4-dihydroxybenzene. Such advance polymerizations require exceptionally pure monomer to ensure high molar mass.

## 4.3 Properties

Polythiazyl $(SN)_x$ is gold colored with metallic luster and electrically conductive. It is a superconductor at very low temperature (below 0.26 K). It is stable in air and insoluble in many solvents. The melting point of polysulfone is more than 500 °C; it is very heat resistant, but poor mechanical properties and difficult to process. In this manner, PAES would be an appropriate choice to overcome these properties. Polysulfones are rigid, high-quality transparent, and its glass transition temperature is in between 190 and 230 °C. Its stiffness and higher strength and dimensional stability are retained in the temperature range from −100 to 150 °C.

Poly(aryl ether sulfone)s are made out of aromatic, ether, and sulfonyl groups. For an examination of the properties of individual constituents, poly(phenylene sulfone) can fill in, for instance, which is comprised of sulfonyl and phenyl groups (thermally stable groups). Poly(phenylene sulfone) has an amazingly high softening temperature (520 °C). In any case, the polymer chains are additionally so inflexible that poly(phenylene sulfone) deteriorates before softening and can subsequently not be thermoplastically handled. Hence, flexible components must be incorporated into the chains; this is done as ether groups. Ether groups permit a free revolution of the polymer chains. This prompts an altogether decreased melting point and furthermore enhances the mechanical properties by an expanded effect quality. The alkyl groups in bisphenol-A demonstrates additionally as a flexible component.

## 4.4 Applications

The electrochemical characteristics of organic compounds in acetonitrile/TEAP solution at $(SN)_x$ paste electrode is reported by Cheek and Horine [35]. The current density formed either by hydroquinone oxidation or by addition of $HClO_4$ to benzoquinone is found to be considerably less than that for this process at platinum or vitreous carbon. A similar effect was observed for proton reduction at (SN), paste, indicating, as for $(SN)_x$ single-crystal electrodes, a high hydrogen overpotential for the electrode surface. Comparative experiments investigating the redox behavior of the benzoquinone/hydroquinone system in aqueous media at $(SN)_x$ paste indicate that slow protonation kinetics are involved in the voltammetric behavior observed in this system. Oxidation of pyrrole produces a black film on the surface of $(SN)_x$ paste, which possesses voltammetric properties similar to those of polypyrrole films that are formed at other electrodes. The dependence of the current observed in the process upon heterocyclic concentration indicates that the process corresponds to shift of the $(SN)_x$ oxidation which is caused by the nucleophilic interaction of the heterocycles with the $(SN)_x$. Further negative shifts in the potential for $(SN)_x$ oxidation are observed in the presence of several alkylpyridines having greater nucleophilicities than that of pyridine itself. The addition of the less nucleophilic compound thiazole did not produce this effect. These results indicate that the oxidation of $(SN)_x$ itself leads to the formation of rather reactive products, while the $(SN)_x$ surface behaves essentially as a noninteracting, metallic electrode in the potential region between +0.95 and −0.40 V versus SCE, over which $(SN)_x$ is electrochemically stable in acetonitrile.

Nowak et al. [36] investigated that the covalent, metallic conductor, that is, $(SN)_x$, was used as an electrode material in the presence of aqueous media. Irreversible redox couples were used as kinetic probes to monitor changes in the $(SN)_x$ electrode surface/solution interface after chemical modification resulting from the strong interaction of $(SN)_x$ with metal cations. The iodate reduction and iodide oxidation reactions permitted the study of two different potential regions. Chromium (III) remained in a cationic form immobilized on parallel $(SN)_x$ surfaces in both the potential regions. The ferro-ferricyanide couple was used to determine the electrochemical parameters and electrode surface areas of $(SN)_x$ polymer [37].

Because of their high reactivity, violated polymers were used for optoelectronics and engineering [38]. Polythiols were used as a building block for the engineering of macromolecular materials [39].

Consumption of Lithium (Li) and sulfur (S) battery requires very less costing and produces high power as compared to lithium battery [40, 41]. By using electrocatalysts as a host material, the performance of lithium and sulfur battery is increased [42, 43]. Huang et al. reported the synthesis of Iron phosphide (FeP)/reduced graphene oxide (rGO)/carbon nanotubes (CNTs) composite by using graphene oxide and DMF through solvothermal and phosphorization process [44, 45].

Lithium-polysulfur (Li-polyS) batteries possess high storage capacity as compared to the lithium-sulfur (Li-S) battery. Chang et al. investigated the use of polysulfur–graphene nanocomposite (poly-SGN)) for the storage capacity of Li-poly batteries [46].

Polysulfone is a good high-performance polymer among all melt-processable thermoplastics because it provides the highest service temperature. Flame retardant ability and high hydrolysis stability are reasonable properties for hemodialysis, wastewater recovery, food and beverage processing, fuel cells, gas separation, and biomedical applications. These polymers are also used in the automotive and electronic industries. It can be reinforced with glass fibers with higher tensile strength and Young's modulus.

## 4.5 Conclusion

Polythiazyl, polythiol, polysulfide, and polysulfone compounds are used in enormous fields varying from adhesives to biomedical. The synthesis of polymeric sulfur nitride done by using powdered tetrasulfur nitride. The synthesis of polythiols can be well achieved by the RAFT polymerization method. The protein immobilization on gold surface and Surface plasmon resonance (SPR) sensor surface using low-cost polythiol surface are successfully achieved.

# References

[1]  Greenwood, N. N., & Earnshaw, A. Chemistry of the elements, 2nd, butterworth-heinemann, 1997, 725–727, ISBN 0080379419.

[2]  Labes, M. M., Love, P., & Nichols, L.F. Polysulfur nitride – a metallic, superconducting polymer, Chem. Rev., 1979, 79(1), 1–15.

[3]  MacDiarmid, A. G., Mikulsk, C. M., Heeger, A. J., Garito, A. F., & Weber, D. C. Polymeric sulfur nitride (Polythiazyl), (SN)$_x$, Inorg. Synth., 1983, 22, 143–149.

[4]  Cohen, M. J., Garito, A. F., Heeger, A. J., MacDiarmid, A. G., Mikulski, C. M., Saran, M. S., & Kleppinger, J. Solid state polymerization of S$_2$N$_2$ to (SN)$_x$", J. Am. Chem. Soc., 1976, 98(13), 3844–3848.

[5]  Ronald, D. A. Inorganic and Organometallic Polymers, John Wiley & Sons, 2001, 264, ISBN 9780471241874.

[6]  Franco, C. New developments in the study of the structure of parathiocyanogen: (SCN)x, Inorg. Polym. J. Inorg. Organomet. Polym., 1997, 7(1), 35–50.

[7]  Levy, G. C., Lichter, R. L., & Nelson, G. L. Carbon-13 nuclear magnetic resonance for organic chemists, John Wiley, New York, 1980.

[8]  Koval, I. V. Synthesis, Structure, and physicochemical characteristics of thiols, J. Organ. Chem., 2005, 41(5), 631–648.

[9]  Hoyle, C. E., Lowe, A. B., & Bowman, C. N. Thiol-click chemistry: A multifaceted toolbox for small molecule and polymer synthesis, Chem. Soc. Rev., 2010, 39(4), 1355–1387.

[10]  Kade, M. J., Burke, D. J., & Hawker, C. J. The power of thiol-ene chemistry, J. Polym. Sci. Part A: Polym. Chem., 2010, 48(4), 743–750.

[11]  Kolb, H. C., Finn, M. G., & Sharpless, K. B. Click chemistry: Diverse chemical function from a few good reactions, Angew. Chem. Int. Ed., 2001, 40(11), 2004–2021.

[12]  Lowe, A. B., Hoyle, C. E., & Bowman, C. N. Thiol-yne click chemistry: A powerful and versatile methodology for materials synthesis, J. Mater. Chem., 2010, 20(23), 4745–4750.

[13]  Ishida, H., Rickert, S. E., Hopfinger, A. J., Lando, J. B., Baer, E., & Koenig, J. L. Epitaxial polymerization of (SN)$_x$: Chemical defects (Article) Epitaxial polymerization of (SN)$_x$, J. Appl. Phys., 1980, 51(10), 5188–5193.

[14]  Rickert, S. E., Ishida, H., Lando, J. B., Koenig, J. L., & Baer, E. Lattice effects on structures and topochemistry, J. Appl. Phys., 1980, 51(10), 5194–5200.

[15]  Akhtar, M., Chiang, C. K., Heeger, A. J., Milliken, J., & MacDiarmid, A. G. Synthesis of metallic polythiazyl halides from tetrasulfur tetranitride, Inorg. Chem., 1978, 17(6), 1539–1542.

[16]  Mikulski, C. M., Russo, P. J., Saran, M. S., MacDiarmid, A. G., Garito, A. F., & Heeger, A. J. Synthesis and structure of metallic polymeric sulfur nitride, (SN)x, and its precursor, disulfur dinitride, S$_2$N$_2$, J. Am. Chem. Soc., 1975, 97(22), 6358–6363.

[17]  Zhang, S., Pattacini, R., & Braunstein, P. Reactions of a phosphinoimino-thiazoline-based metallo-ligand with organic and inorganic electrophiles and metal-induced ligand rearrangements, Organometallics, 2010, 29(24), 6660–6667.

[18]  Sonnenschein, L., & Seubert, A. Separation of inorganic anions using a series of sulfobetaine exchangers, J. Chromatogr. A., 2011, 1218(8), 1185–1194.

[19]  Cataldo, F., & Fiordiponti, P. Possible structure of parathiocyanogen—II. electrochemical synthesis, $^{13}$C NMR and conductivity measurements on undoped and iodine doped samples, Polyhedron, 1993, 12(3), 279–284.

[20]  Rostislav, L., Galagan, L. P., Shepetun, V. A., Litvin, B., & Minaev, F. Synthesis, and spectroscopic characterization, of a new (aryl-SCN)n polymer: Polythiocyanatohydroquinone, Gleb V. Baryshnikov, J. Mol. Struct., 2015, 1096, 15–20.

[21] Jones, W. H. Metastable polymers of the nitrogen oxides. 1. Open chain nitric oxide analogues of polythiazyl: A MNDO/AM1 study, J. Phys. Chem., 1991, 95(6), 2588–2595.

[22] Jones, W.H. Metastable polymers of the nitrogen oxides. 2 Open-chain polymers of the nitric oxide dimers and of nitrous oxide: A MNDO/AM1 study, J. Phys. Chem., 1992, 96(2), 594–603.

[23] Ancelin, H., Hauptman, Z. V., Banister, A. J., & Yarwood, J. A Fourier-transform infrared spectroscopic study of the surface hydrolysis of polythiazyl, J. Polym. Sci. B Polym. Phys., 1990, 28, 1611–1619.

[24] Ramakrishnan, S., & Chien,, J. C. W. Phenyl thiazyl compounds: Synthesis and electrical conductivity, J. Polym. Sci. Part A: Polym. Chem., 1987, 25(5), 1433–1443.

[25] Morgane, L. N., & Renaud, N. Polythiol copolymers with precise architectures: A platform for functional materials, Polym. Chem., 2014, 5(16), 4601–4611.

[26] Nicolay, R. Synthesis of well-defined polythiol copolymers by RAFT polymerization, Macromolecules, 2012, 45(2), 821–827.

[27] Tongxin, W., Rajagopalan, S., Alexandru, K., Andy, H. T., Kyethann, F., James, M., & Paul, C. W. Synthesis of amphiphilic triblock copolymers as multidentate ligands for biocompatible coating of quantum dots, Colloids Surf. A: Physicochem. Eng. Aspects, 2011, 375(1–3), 147–155.

[28] Kenichi, A., Ryota, I., & Masatsugu, Y. Novel dendritic polyenes for application to tailor-made thiol-enephotopolymers with excellent UV-curing performance, Prog. Org. Coat., 2016, 100(c), 105–110.

[29] Qi, F., Jianhua, L., & Wenfang, S. Preparation and photopolymerization behavior of multifunctional thiol–Ene systems based on hyperbranched aliphatic polyesters, Prog. Org. Coat., 2008, 63(1), 100–109.

[30] Nikolaos, P., Petri, K., Emma, M., Hannu, I., & Jukka, S. Novel thiol- amine- and amino acid functional xylan derivatives synthesized by thiol–Ene reaction, Carbohydr. Polym., 2015, 131, 392–398.

[31] Yu, T., Yu, C., Yuxiao, T., Zhenguo, Y., & Haiping, W. Self-healing isotropical conductive adhesives filled with Ag nanowires, Mater. Chem. Phys., 2014, 148(3), 778–782.

[32] Yan, C. Y., Min, Z. R., & Ming, Q. Z. Preparation and characterization of microencapsulated polythiol, Polymer, 2008, 49(10), 2531–2541.

[33] Falk, M., Anne, F., Eric, M., Markus, R., Stefan, L., Hans-Joachim, K., & Wolfgang, E. Synthesis of nanoparticle/ligand composite thin films by sequential ligand self-assembly and surface complex reduction, J. Coll. Interface Sci., 2013, 389(1), 23–30.

[34] Enrique, J. C., Guillermo, R. R., José, M. H. M., & Michael, L. Polymethacrylate monoliths with immobilized poly-3-mercaptopropyl methylsiloxane film for high-coverage surface functionalization by thiol-ene click reaction, J. Chromatogr. A, 2014, 1367, 123–130.

[35] Cheek, G., & Horine, P. A. Electrochemical behavior of organic compounds at the $(Sn)_x$ paste electrode, J. Electrochem. Soc., 1985, 132(1), 115–119.

[36] Nowak, R. J., Kutner, W., Voulgaropoulos, A., Rubinson, J. F., Mark, H. B., & Macdiarmid, A. G. The polythiazyl, $(Sn)_x$, electrode: Surface modification with metal cations, J. Electrochem. Soc., 1981, 128(9), 1927–1931.

[37] Nowak, R. J., Kutner, W., & Mark, H. B. Behavior of polymeric sulfur nitride, $(SN)_x$, Electrodes in aqueous media, J. Electrochem. Soc., 1978, 125(2), 232–240.

[38] Stenzel, M. H. Bioconjugation using thiols: Old chemistry rediscovered to connect polymers with nature's building blocks, ACS Macro Lett., 2013, 2(1), 14–18.

[39] Ma, X., Zhou, Z., Jin, E., Sun, Q., Zhang, B., Tang, J., & Shen, Y. Facile synthesis of polyester dendrimers as drug delivery carriers, Macromolecules, 2013, 46(1), 37–42.

[40] Seh, Z.W., Sun, Y., Zhang, Q., & Cui, Y. Designing high-energy lithium-sulfur batteries, Chem. Soc. Rev., 2016, 45(20), 5605–5634.

[41]  Peng, H. J., Huang, J. Q., & Zhang, Q. A review of flexible lithium–Sulfur and analogous alkali metal–Chalcogen rechargeable batteries, Chem. Soc. Rev., 2017, 46(17), 5237–5288.

[42]  Yuan, H., Chen, X., Zhou, G., Zhang, W., Luo, J., Huang, H., Gan, Y., Liang, C., Xia, Y., & Zhang, J. Efficient activation of $Li_2S$ by transition metal phosphides nanoparticles for highly stable lithium-sulfur batteries, ACS Energy Lett., 2017, 2(7), 1711–1719.

[43]  Fang, R., Zhao, S., Sun, Z., Wang, D. W., Amal, R., Wang, S., Cheng, H. M., & Li, F. Polysulfide immobilization and conversion on a conductive polar MoC @ $MoO_x$ material for lithium-sulfur batteries, Energy Storage Mater, 2018, 10, 56–61.

[44]  Huang, S., Lim, V. Y., Zhang, X., Wang, Ye., Zheng, Y., Kong, D., Ding, M., Yang, A. S., & Yang, Y. H. Regulating the polysulfide redox conversion by iron phosphide nanocrystals for high-rate and ultrastable lithium-sulfur, Nanoenergy, 2018, 51, 340–348.

[45]  Chung, H. S., Luo, L., & Manthiram, A. $TiS_2$-polysulfide hybrid cathode with high sulfur loading and low electrolyte consumption for lithium-sulfur batteries, ACS Energy Lett., 2018, 3(3), 568–573.

[46]  Chang, H. C., & Manthiram, A. Covalently-grafted polysulfur-graphene nanocomposites for ultrahigh sulfur-loading lithium-polysulfur batteries, ACS Energy Lett., 2017, 3(1), 72–77.

# 5 Organometallic polymers: ferrocene and photo-cross-linkable polymers

**Abstract:** In this chapter, various synthetic approaches to ferrocene (Fc) and photo-cross-linkable are discussed in detail. For the synthesis of the main chain Fc-based polymers, polycondensation reaction, ring-opening polymerization, and other synthetic methods have been summarized. Photo-cross-linkable polymers and ultraviolet-curable resins have significant applications such as coatings, adhesives, photoresists, and printing plates. Some potential applications of these monomers and polymers are also described.

**Keywords:** synthetic, photopolymerization, polycondensation, application

## 5.1 Introduction

In 1951, ferrocene (Fc) was discovered [1, 2] and its exact structure was explicated by Wilkinson and coworkers [3]. The similarity of its reactivity to benzene motivated the scientists to name this new iron-sandwiched compound as Fc. The structure disclosure of Fc was a breakthrough in the world of chemistry and steered to the birth of recent organometallic chemistry. Fc rapidly attracted thoughtfulness of technical communities and the scientist on expenses of its charming chemistry. Scientists started to develop synthetic approaches using Fc derivatives, as well as discovered the uses in an extensive array of scientific zones. From 1984 to till now numerous categories of ferrocenyl compounds have been produced and assessed for their glucose-sensing applications.

Fc was synthesized from cyclopentadiene by using metal–hydrogen exchange method and further followed by the reaction with Fe(II) chloride. Inherent torsional freedom cyclopentadiene moiety around the Fe atom is easily available in the polymer. Vinyl Fc produces both homo- and copolymer by using free radical polymerization. In the initiation step, the electron donation from iron atoms generates growing radicals into its anion form, which results in the termination step. The $Fe^{2+}$ centers easily rearrange to form a paramagnetic and ionic bonded Fe(III) species.

As a result, extensive chain transfer and the formation of the branched structure occurred. The molecular weight of these polymers is found to be about 2,50,000 Da. That is the only reason for the insolubility of these polymers. In condensation polymerization, various functional Fc is used to form functional polymers. As a result of condensation polymerization, medium- or low-molecular-weight (50,000 DA) polymers with high polydispersity indices are synthesized. In ring-opening polymerization, low-molecular-weight polymers are formed.

https://doi.org/10.1515/9781501514609-006

Chen et al. reported the use of stimulus-responsive polymer brushes for controlling the surface wettability and they also act as a thin film sensor [4]. Sun et al. reported the use of film for the wettability of the surface [5]. Pei et al. reported that the film is used for controlling wetting of the surface by electrochemical and its interaction with the living cells [6].

Isaksson and coworkers investigated that the electronic wettability switch devices such as dodecyl benzene sulfonic acid were used for knowing the wettability of the surface [7]. Whittell and Manners reported the use of Fc-containing polymers for making optical and electronic devices [8]. Abbott and Whitesides reported the use of Fc for controlling the wettability of the surface [9].

Min and Chang studied about the preparation of polyazomethine (PAM) nanocomposites by using organoclay $C_{12}$-Montmorillonites (MMT) through in situ interlayer polymerization [10]. Iwan et al. studied the use of aromatic PAMs for making photovoltaic devices [11]. Senel et al. investigated that the biosensor is prepared by using horseradish peroxidase through free-radical copolymerization [12]. Soldatkin et al. studied the preparation of biosensor [13]. Silvana et al. studied the use of a creative based biosensor for monitoring of hemodialysis [14].

Dang et al. studied the use of conjugated polymers for the preparation of photovoltaic and lightweight devices [15]. Kim et al. studied the use of bromine-functionalized poly(3-hexylthiophene) in the packing of conjugated polymers and making bulk heterojunction photovoltaics [16]. Contoret et al. reported the use of charge-transporting semiconductor material for making organic field effect transistor [17].

Charpoy reported the use of Liquid crystal polymers (LCPs) in the preparation of optical and electrical devices [18]. Ravikrishnan et al. studied the use of Liquid crystal polymers (LCPs) for arranging the mesogenic groups into the ordered structure of mesophase [19]. David et al. reported that the photo-cross-linkable polymers are used in the preparation of photoconductors, display and optical storage devices, LC elastomers, and LC thermosets [20]. Ravikrishnan et al. studied that the flexibility of photo-cross-linkable LCPs is affected by the methylene groups [21].

## 5.2 Synthetic methods

Different types of ring-opening mechanism are shown in Figure 5.1.

### 5.2.1 Synthesis of Fc-containing polymers

#### 5.2.1.1 Fc-based polypyrrole
Functionalized polypyrrole dendrimers combined electronic conductivity of the conducting polymer with the redox properties of Fc to mediate electron transfer

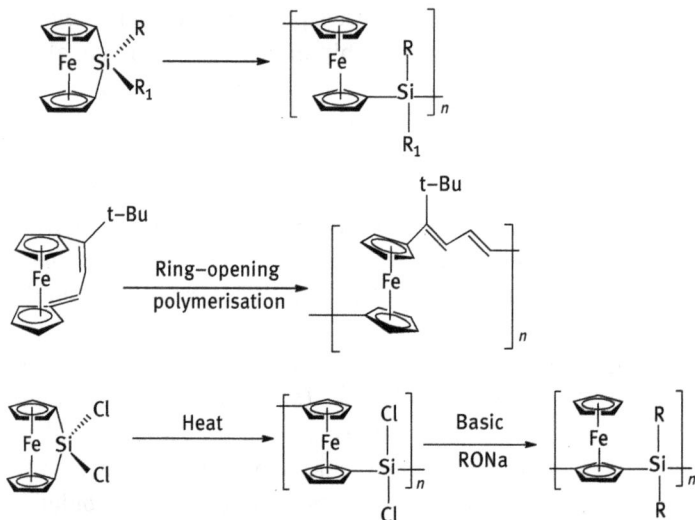

**Figure 5.1:** Ring-opening polymerization via various approaches.

reactions and improved sensing behavior. Palomera et al. reported the synthesis of mediator-less copolymer (electrochemical biosensor) of pyrrole and ferrocene carboxylate-modified pyrrole by electropolymerization method [22]. Furthermore, Soon et al. reported the synthesis of Fc-based pyrrole derivatives with differing spacer linkages by coupling reactions (Figure 5.2b) [23]. Functionalized polypyrrole materials are fairly easy to deposit from nonaqueous electrolytes and have potential applications in electronic or perhaps sensor devices. Senel reported the synthesis of the copolymer (Fc-branched polypyrrole) by condensation between Fc ethanol, 3-(1H-pyrrol-1-yl) propanoic acid, and 3-(1H-pyrrol-1-yl and ferrocene-pyrrole) propanoic acid(Figure 5.2c) [24]. With the use of Fc-based pyrrole derivatives, it was discovered that film growth using potential sweeping was the most appropriate for adherent film formation with well-defined redox electrochemistry and surface-confined behavior. Moreover, an advantageous feature of these functionalized polypyrrole materials was not hard to deposit from nonaqueous electrolytes and have potential applications in electronic or perhaps sensor devices.

### 5.2.1.2 Fc-based chitosan

Fc-based chitosan (CHIT) derivatives in the aldehyde group of Fc can just react with the $NH_2$ group of CHIT via a Schiff base and succeeding reduction by $NaCNBH_3$ [25]. CHIT-Fc/GOD film electrodes are extremely vulnerable in the electroanalysis of

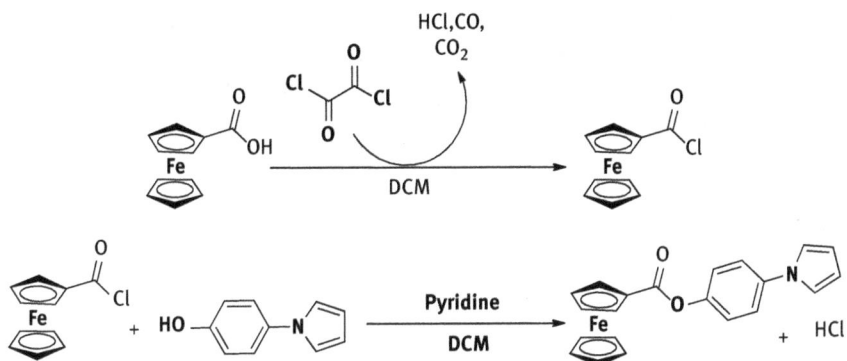

(a) (6-[4-(1H-pyrrol-1-yl]phenoxy)hexyl)ferrocene

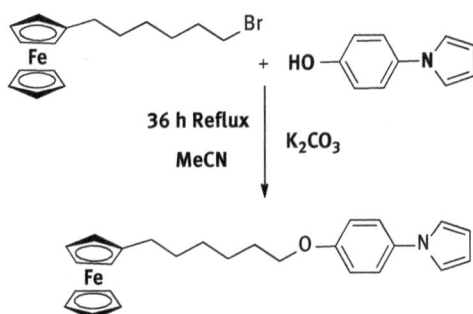

(b) (6-[4-(1H-pyrrol-1-yl]phenoxy)hexyl) ferrocene (1)

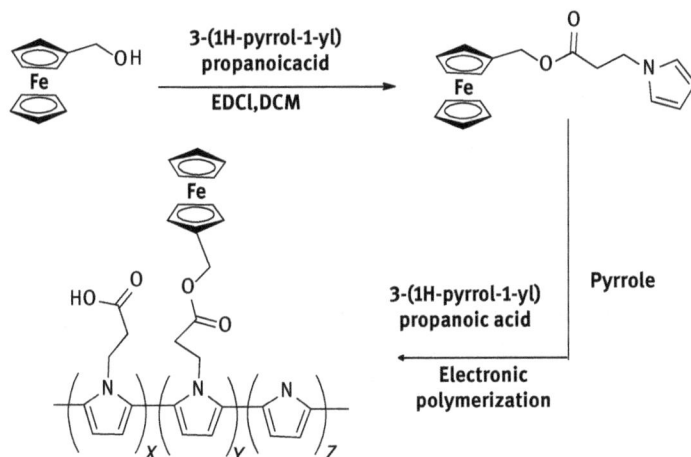

(C) Co -polymer(py/py-Fc/py-COOH)

**Figure 5.2:** Preparation of (a) (6-[4-(1H-pyrrol-1-yl]phenoxy) hexyl) ferrocene, (b) ([4-(1H-pyrrol-1-yl) phenoxy] carbonyl) ferrocene, and (c) poly(pyrrol/pyrrol-NH2/pyrrol-Fc).

glucose, but as a result of the hydrophobic dynamics of the ferrocenyl, the quantity of redox active websites in this particular polymer is actually restricted. That issue was overcome by functionalized polysiloxane with a CHIT composite of Fc in which uniform dispersal of Fc in the stability and CHIT matrix between hydrophilic/hydrophobic properties of nanocomposite made smooth mass transporting redox movie and then showed a greater transport rate in aqueous condition [26].

### 5.2.1.3 Fc-based cyclodextrin

Practical application of natural CD (cyclodextrin; a, b, and g CDs) is restricted because of poor aqueous solubility, especially that of b CD. The improved aqueous solubility of Fc makes it a great addition for hydrophobic holes of CD in aqueous media. An Fc-carbonyl-b-CD inclusion complex was prepared by highly soluble carbonyl-b-CD, where Fc was found within the gap of carbonyl-b-CD [27]. The carbon nanotube incorporation in Fc-carbonyl-b-CD inclusion complex boosted electron transfer between the electrode surface and FcCD complex to create an amperometric glucose biosensor [28]. Molecular size selective glucose sensor was also prepared from thiolated a CD in which glucose molecules fit directly into the template molecules as well as competed nicely with the electrochemically active compound for insertion complex formation [29].

*Grafting of polyvinylferrocenes tetraethoxysilane (PVFc-TEOS):* Argon, toluene, and PVFc-TEOS are mixed in the flask to prepare a solution. This solution is continuously stirred for 30 h at 200 °C. The wafer is dissolved in tetrahydrofuran (THF) to remove the polymer.

### 5.2.1.4 Synthesis of Fc PAMs

*Synthesis of Fc-PAM-A:* 1-(m-Formylphenyl)-1'-(5-formyl-2-methoxphenyl) Fc, propane-1,3-diamine is dissolved in absolute EtOH and then also add few drops of piperidine to it, to get a reaction mixture. The reaction is carried out in the three-necked flask to get an orange precipitate of product.

*Synthesis of Fc-PAM–B:* 1-(m-Formylphenyl)-1'-(5-formyl-2-methoxphenyl) Fc, decane-1,10-diamine is dissolved in absolute EtOH and then also add few drops of piperidine to it, to get a reaction mixture. The reaction is carried out in the three-necked flask to get an orange film of product.

A band at 1,648 cm$^{-1}$ is attributed due to CH = N stretching. Thermogravimetric analysis (TGA) curves of the Fc-PAMs A and B show loss in weight till 250 °C. The initial decomposition of these polymers was known as polymer decomposition temperature [30, 31].

### 5.2.2 Synthesis of composite films

#### 5.2.2.1 Fc-doped vanadium pentoxide (VXG)

Fc and isopore polycarbonate membranes are used. To prepare a mixture, $FeCp_2$ is dissolved in acetonitrile and then add $V_2O_5$ xerogel to it [32]. The obtained precipitate is filtered and then dried. A green-blue powder is obtained as a product.

#### 5.2.2.2 Glucose oxidase-type X-S – photo-cross-linkable polyvinyl alcohol–styrylpyridinium residues ($GO_x$–PVA–SbQ) and $FeCp_2$–VXG composite films

An enzymatic solution is prepared by dissolving $GO_x$ in a buffer solution having pH 7. This enzymatic solution is added to PVA–SbQ and $FeCp_2$–VXG mixture. This mixture is spread and then dried for 24 h. The film is irradiated and covered with polycarbonate membrane.

   A singlet is observed for ferricinium cations by using $^{57}$Fe Mossbauer spectroscopy in the $FeCp_2$–VXG spectrum [33]. A doublet of $Fe^{3+}$ is observed in $^{57}$Fe Mossbauer spectrum of $FeCp_2$–VXG due to distorted ferricinium ions [34].

### 5.2.3 Photo-cross-linkable polyfluorenes for light-emitting diode

Polyfluorenes are very widely used polymers as blue light-emitting polymer because of their reasonably good electro- and photoluminescence efficiencies.

#### 5.2.3.1 Synthesis of 2-[9,9-di[4-(hexyloxy)phenyl]-7-(4,4,5,5-tetramethyl-1,3, 2-dioxaborolan-2-yl)-9H-2-fluorenyl]-4,4,5,5-tetramethyl-1,3, 2-dioxaborolane (M1)

2,7-Dibromo-9,9-bis(4-hydroxyphenyl)] fluorine is dissolved in dry THF and add *n*-butyllithium/hexane dropwise to it. After 20 min add 2-isopropoxy-4,4,5,5-tetramethyl-1,3,2-dioxaborolane to the mixture and is continuously stirred for 28 h. The obtained product is white crystalline solid. $^1$H NMR (ppm): peak appears at 7.75 due to the aromatic ring, the peak appears at 1.65 due to $OArCH_2A(CH_2)_4ArCH_3$ at meta position, and the peak appears at 1.24 due to $CH_3$.

#### 5.2.3.2 3-({[6-(4-{3,6-Dibromo-9-[4-({6-[(3-methyl-3-oxetanyl)methoxy]hexyl}oxy) phenyl]-9H-9-fluorenyl}-phenoxy)hexyl]methyl}-3-methyloxetane (M2)

When potassium hydride and potassium iodide are dissolved in acetonitrile, a solution is obtained. M1 and 3-(((6-Bromohexyl)oxy)methyl)-3-methyloxetane are added slowly to the solution. The obtained product is a white, colorless oil. $^1$H NMR (ppm): peak appears at 6.85 due to aromatic protons at meta position, the peak appears at 4.25 due to ring protons, peak appears at 3.45 due to $ArCH_2ArOArCH_2$ at meta position, and the peak appears at 0.83 due to $ArCH_3$

### 5.2.3.3 Polyfluorene with blue, green, and red (PFB, PFG, and PFR)

PFB, PFG, and PFR are prepared in the presence of a palladium catalyst through Suzuki-coupling reaction and they emit blue, green, and red light, respectively (Figure 5.3) [35].

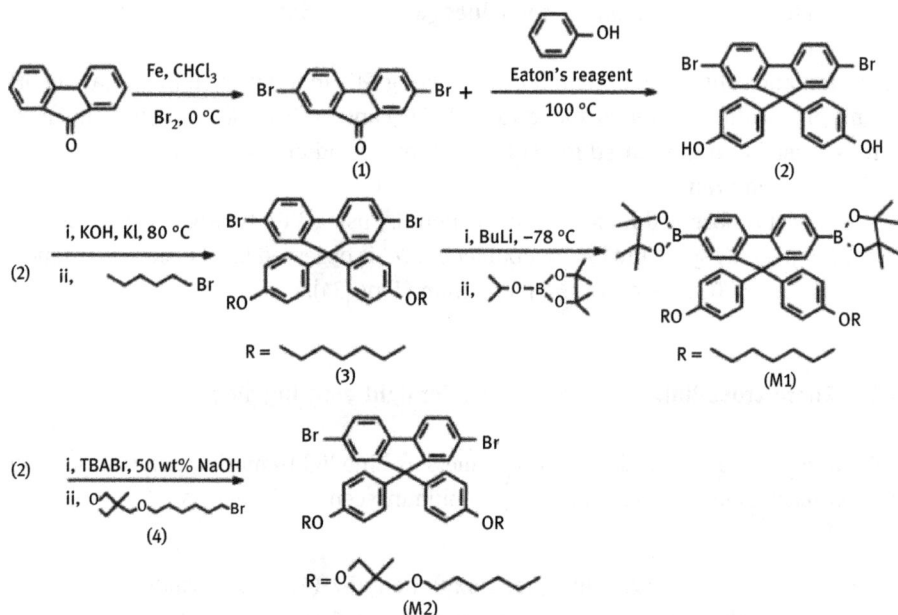

Figure 5.3: Outlines the synthesis of the monomers for photo-cross-linking [35].

Synthesis of PFB is as follows. $M_2$, $M_1$, and $M_3$ are dissolved in toluene in two-necked round flask to prepare a solution. The obtained polymer is a light green solid. Similarly, PFG and PFR are prepared by varying the ratio of $M_1$, $M_2$, and $M_3$ (Figure 5.4).

### 5.2.4 Photo-cross-linkable acrylate

Photo-cross-linkable acrylate polymers are synthesized by using his (diarylamino) biphenyl acrylate and cinnamate acrylate as monomers and AIBN is used as an initiator. This reaction occurs in the presence of palladium as a catalyst. This reaction is a photochemically induced 2 + 2 cycloaddition reaction [36]. The obtained residue is dissolved in THF, filtered, and then dried. The obtained product is a white

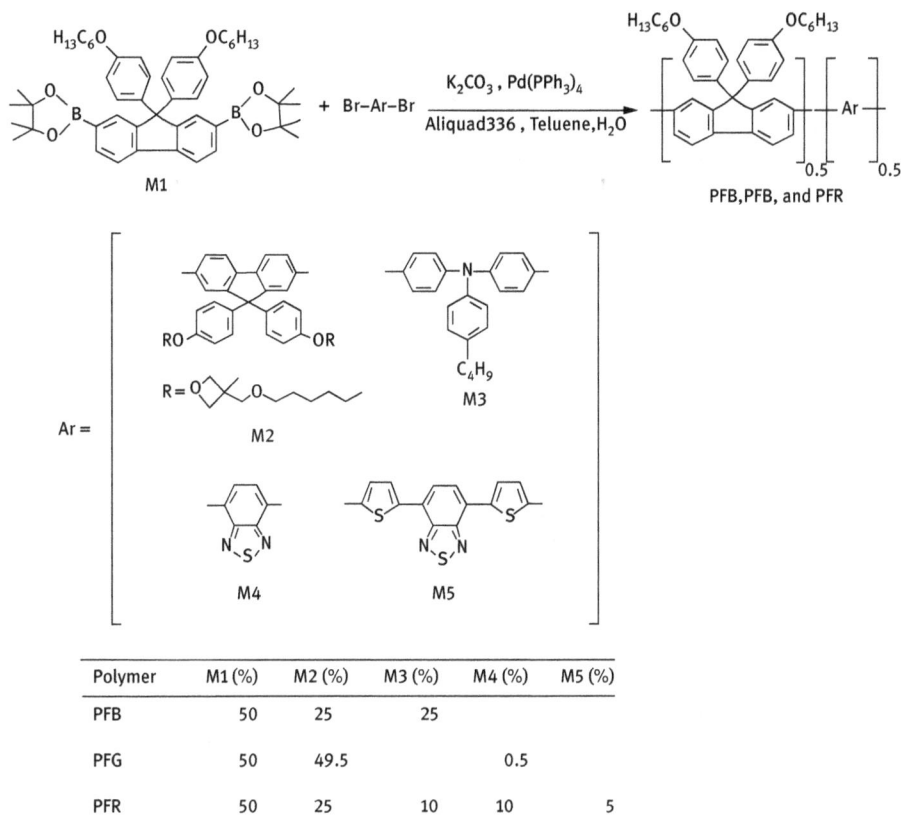

**Figure 5.4:** Synthesis of PFB, PFG, and PFR [35].

| Polymer | M1 (%) | M2 (%) | M3 (%) | M4 (%) | M5 (%) |
|---------|--------|--------|--------|--------|--------|
| PFB | 50 | 25 | 25 | | |
| PFG | 50 | 49.5 | | 0.5 | |
| PFR | 50 | 25 | 10 | 10 | 5 |

powder. As the ionization potential of diode increases, the efficiency of light-emitting diode improves. The device having polymer P3 shows high efficiency [37].

### 5.2.5 Photo-cross-linkable π-conjugated cruciform molecules for optoelectronic application

#### 5.2.5.1 Synthesis of photo-cross-linkable π-conjugated cruciform molecules

Photo-cross-linkable p-type thiophene-based cruciform semiconductor molecule is synthesized by using photoreactive end-groups, acrylate and 1,4-pentadien-3-yl group as monomers and AIBN is used as an initiator. This reaction is a photochemically induced 2 + 2 cycloaddition reaction. FT-IR confirms that photopolymerization process occurs. The absorption bands are observed at 1,670 and 1,743 $cm^{-1}$, due to the presence of diene or acrylate photoreactive groups. The bands are observed at 1,740 $cm^{-1}$. Shifting of the C = O group occurs due to cross-linking.

## 5.2.6 Synthesis of photo-cross-linkable liquid crystalline polymer

### 5.2.6.1 Synthesis of 2, 6-bis(4-hydroxybenzylidene)cyclohexanone (BHBCH)

Cyclohexanone, triethylamine (TEA), thionyl chloride, and 2-chloroethanol are used. Cyclohexanone and $p$-hydroxybenzaldehyde are dissolved in absolute ethanol and then add HCl. It is washed with ethanol to give a yellowish green product [38–40].

### 5.2.6.2 Synthesis of 4-phenyl-2,6-bis(4-chlorocarbonylphenyl)-pyridine

When (4-phenyl)-2,6-bis(4-carboxyphenyl)-pyridine is dissolved in dry benzene, a solution is obtained. The reaction occurs in a round-bottomed flask equipped with a reflux condenser. Add thionyl chloride and DMF dropwise to the solution. The obtained product is a yellow powder [41].

### 5.2.6.3 Synthesis of poly{2,6-bis[4-(2-ethyloxy)benzylidene] cyclohexanaone-
###          (4-phenyl)-2,6-bis(4-phenyl)pyridine dicarboxylate}

When 2,6-bis[4-($m$-hydroxyalkyloxy)benzylidene] cyclohexanone is dissolved in dry THF, a solution is obtained. Add TEA to the solution with constant stirring in the presence of an inert atmosphere. The obtained product is a yellow precipitate. The monomers and polymers with spacers having two and four methylene units do not show any LC phases, whereas the polymers with longer spacers ($m$ = 6, 8 and 10) show homogeneous nematic LC phase. This is observed because of the flexible methylene spacer group of polymers III–V which facilitates the LC phases [42]. The presence of bulky heterocyclic ring system in the polymer main chain prevents the higher order LC textures like threaded nematic and smectic. Monomer and polymers exhibited grainy-type nematic phases.

## 5.2.7 Synthesis of Fc-containing polyamides and polyurea

Fc-containing polyamides and polyurea can be prepared using varieties of reagents on Fc by intrafacial polymerization (Figures 5.5 and 5.6) [43].

# 5.3 Properties

The glass transition temperature of Fc-based polymers is actually lower than some other neutral or perhaps cationic polymers. Liquid crystal polymers can also be studied by differential scanning calorimetry and TGA. Although an assortment of metallopolymers possesses interesting chemical, electrical, and magnetic properties, the researchers in the field of material sciences focused on metallocene-containing polymers due to their interesting properties that come up from their

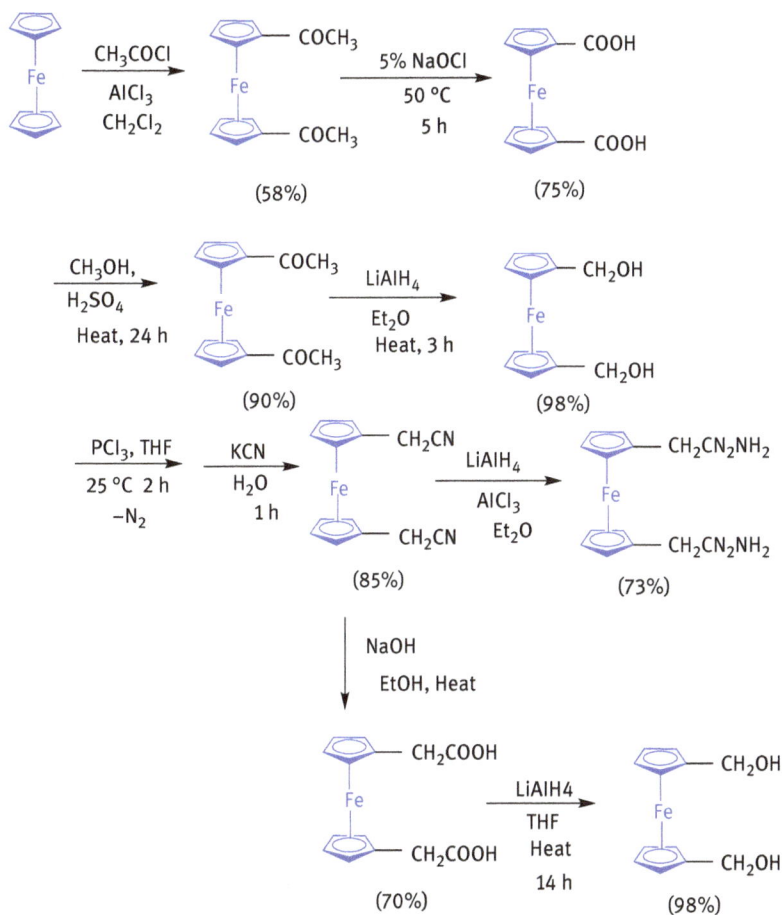

**Figure 5.5:** Synthesis of 1,l'-bis(P-hydroxyethy1)ferrocene.

unique sandwich-like structures including thermal stability and redox behavior. Fc-containing polymers appear to be most common organometallic redox systems due to their electrochemical properties, like electron donating ability, superfast electrochemical response, thermal stability, and redox reversibility which nominate them to be a standard reference system in the electrochemical analysis and cathodic or anodic electrode for energy storage applications. The charge densities, electronic efficiency, cyclability, as well as the charge/discharge ability aren't just the complete requirements for the future electrode materials, but safety, nontoxicity, recyclability, and resource availability are actually needed. Among electroactive materials, Fc and also ferrocenyl-based polymers represent undeniably one of the most promising materials.

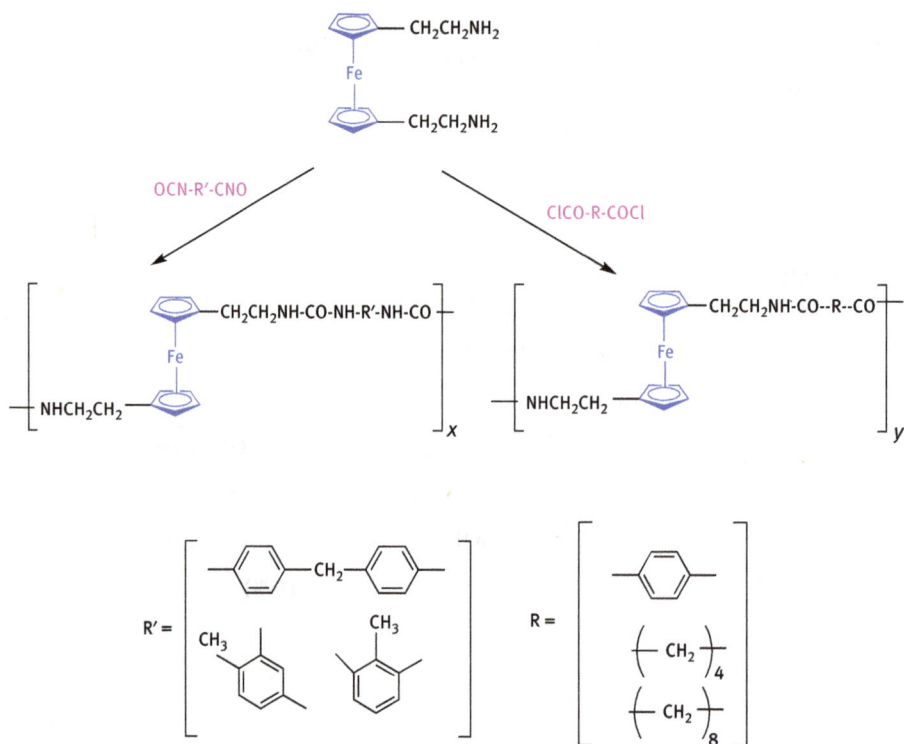

**Figure 5.6:** Synthesis of ferrocene-containing polyamides and polyurea.

## 5.4 Applications

Fc-based polymers have a wide application prospect in different fields like electro-chemistry, biomedicine, optics and sensor and amperometric biosensor, and light-emitting diode due to the unique structures and properties. Fc-based polymers are given in Table 5.1.

## 5.5 Conclusion

Much attention has been paid to the recovery or perhaps recycling of cross-linked polymers due to environmental regulations. To this end, photo-cross-linkable poly-mers with degradable properties have been studied. Metal-containing polymers like polymetallocenes are actually one of the most widely studied organometallic poly-mers. A wide range of transition metal ions has been incorporated into these kinds of structures and even heterobimetallic species have been reported. The organome-tallics containing ligands with conjugated and delocalized π systems will provide

**Table 5.1:** Recent advance and applications of ferrocene-based polymers.

| S. no. | Synthetic approach | Application | Reference number |
|---|---|---|---|
| 1 | Palladium-catalyzed synthesis of ferrocene-based conjugated microporous polymers | Excellent magnetic properties | [44] |
| 2 | Synthesis of polynickelocene using ring-opening polymerization (ROP) | Enhanced magnetic properties | [45] |
| 3 | Electropolymerized 4-(2,5-di (thiophen-2-yl)-1H-pyrrol-1-yl)aniline monomer (SNS-aniline) on pencil graphite electrode, then modifying the polymer-coated electrode surface with di-amino-ferrocene as the mediator, and lastly urease enzyme through glutaraldehyde cross-linking | Bio-electrode for excellent analytical performance | [46] |
| 4 | Hyperbranched ferrocene-containing polysiloxane with 1,1-bis (dimethylvinylsilyl)ferrocene upon pyrolysis produce ceramics | For electromagnetic wave absorption and shielding applications. | [47] |
| 5 | Norbornene-based monomer bearing a nickel (II) complex of Goedken's macrocycle with Grubbs' third generation, resulting in the polymer | – | [48] |
| 6 | Grafted polymers onto polyvinyl chloride using aldehyde and ferrocene carboxyl hydrazide groups as pendant | – | [49] |
| 7 | Azo-functionalized polyesters prepared by solution polycondensation using diacylchloride, azo, and ferrocene dicarboxylic acid | – | [50] |
| 8 | The ferrocenyl tetracyanobutadiene derivative-based conjugated polymer | Solar cell | [51] |
| 9 | Two ferrocene-based hyperbranched polytriazoles prepared by click polymerization using Cu(PPh$_3$)$_3$Br as catalyst | – | [52] |

**Table 5.1** (continued)

| S. no. | Synthetic approach | Application | Reference number |
|---|---|---|---|
| 10 | Redux-responsive poly-(ferrocenylsilane)-based poly (ionic liquid)s and poly(acrylic acid) (PAA) | Act as a reference electrode for possible applications in free impedance sensing, redox-controlled gating, or molecular separations | [53] |
| 11 | Surface-initiated atom transfer radical polymerization (ATRP) used for ferrocene-based brushes | – | [54] |
| 12 | Ferrocene-poly(*para*-phenylene) with π staking interaction | Electronic applications | [55] |
| 13 | *N*-Ferrocenylsulfonyl-2-methylaziridine (fcMAz) synthesized from ferrocene in three steps by anionic polymerization | – | [56] |
| 14 | Ferrocene-based ligand bearing phosphinic groups (Fc (PHOOH)$_2$ = 1,1′-ferrocenediyl-bis(H-phosphinic acid)) formed coordination polymers | – | [57] |
| 15 | Hyperbranched polyurethane-based ferrocene with cytochrome c (cyt c) | – | [58] |
| 16 | Electrospinning technique employed for ferrocenyl-substituted pyrazolines with chitosan and polyethylene terephthalate | Biomedical application | [59] |
| 17 | Hybrid polymer based on poly (*p*-phenylene) modified with ferrocene groups as side chains | DNA sensing | [60] |
| 18 | Organometallic copolymers based on ferrocene and triphenylamine pendants | Cathode active materials for battery | [61] |
| 19 | Conjugated ferrocene-containing poly (fluorenylethynylene)s with triphenylamine, carbazole, or thiophene moieties using Sonogashira coupling reaction | Memory applications | [62] |

**Table 5.1** (continued)

| S. no. | Synthetic approach | Application | Reference number |
|---|---|---|---|
| 20 | Polypropylene (iPP) nanocomposites with carbon-based fillers such as carbon nanotubes using ferrocene as catalyst | – | [63] |
| 21 | Ferrocene-based thiophene as 2-(thiophen-3-yl) ethyl ferrocenoate and 2-(thiophen-3-yl) methyl ferrocenoate copolymerized with 3-hexylthiophene (3HT) | Electronic applications | [64] |
| 22 | β-Cyclodextrin (β-CD) and ferrocene (Fc) introduced int o the main chain of PAA | Supramolecular polymer | [65] |
| 23 | Ferrocene methanol-based materials ferrocenylmethoxy)ethyl methacrylate, 3-(ferrocenylmethoxy) propyl methacrylate, and 4-(ferrocenylmethoxy)butyl methacrylate polymerized by a free radical mechanism | Functional polymers | [66] |
| 24 | Ferrocene-based mixed-metal polymeric microspheres | Hydrogen storage | [67] |
| 25 | Poly(fluorenylethynylene)s with ferrocene moieties | Cathode-active materials for rechargeable lithium batteries | [68] |
| 26 | ATRP for ferrocene-derived polymers, further interacted with multiwalled carbon nanotubes (MWCNTs) | Supramolecular polymer | [69] |
| 27 | Poly(azomethine)ester having ferrocene moieties | Functional materials | [70] |
| 28 | ATRP technique for copolymers with ferrocene moieties | Drug delivery applications | [71] |
| 29 | A composite based on poly(vinyl ferrocene), gelatin, and MWCNT | Biosensor | [72] |
| 30 | Ferrocenyldithiophosphonate functional conducting polymer | Biosensor | [73] |

**Table 5.1** (continued)

| S. no. | Synthetic approach | Application | Reference number |
|---|---|---|---|
| 31 | The ferrocene-based nanoporous organic polymer prepared by coupling 1,1′-ferrocene-dicarboxaldehyde with melamine | Energy application | [74] |
| 32 | Chemical oxidative polymerization employed for conducting polymers having ferrocene moieties | Functional material | [75] |
| 33 | Ferrocene-conjugated copper(II) complexes [Cu(Fc-aa)(aip)](ClO$_4$) and [Cu(Fc-aa)(pyip)](ClO$_4$) of l-amino acid-reduced Schiff bases (Fc-aa), 2-(9-anthryl)-1H-imidaz [1, 10] phenanthroline (aip) and 2-(1-pyrenyl)-1H-imidazo [1, 10] phenanthroline (pyip) used to form coordination polymer (Fc = ferrocenyl moiety) | – | [76] |
| 34 | Poly(ferrocenyl-methylsilane) and its derivatives synthesized by ROP | Energy storage | [77] |
| 35 | Host–guest approach for ferrocene-containing pluronic F127 and β-cyclodextrin linear polymer | Supramolecular chemistry | [78] |
| 36 | Ferrocene-based metalloligand for coordination polymer | Hydrogen uptake capacity | [79] |

applications that are beneficial for the synthesis of new polyunsaturated organic substrates for electronic conductivity and nonlinear optical properties.

# References

[1]  Kealy, T.J., & Pauso, P.L. A new type of organo-iron compound, Nature, 1951, 168, 1039–1040.
[2]  Miller, S.A., & Tebbot, J.A. Tremaine, J.F.Dicyclopentadienyl Iro, J. Chem. Soc., 1952, 632–635.
[3]  Wilkinson, G., Rosenblum, M., Whiting, M.C., & Woodward, R.B. the structure of iron bis-cyclopentadienyl, J. Am. Chem. Soc., 1952, 74, 2152–2126.
[4]  Chen, T., Ferris, R., Zhang, J., Ducker, R., & Zauscher, S. Stimulus-responsive polymer brushes on surfaces: Transduction mechanisms and applications, Prog. Polym. Sci., 2010, 35, 94–112.

[5]   Sun, W., Zhou, S., You, B., & Wu, L. A facile method for the fabrication of superhydrophobic films with multiresponsive and reversibly tunable wettability, J. Mater. Chem. A, 2013, 1, 3146–3154.

[6]   Pei, Y., Travas-Sejdic, J., & Williams, D.E. Reversible electrochemical switching of polymer brushes grafted onto conducting polymer films, Langmuir, 2012, 28, 8072–8083.

[7]   Isaksson, J., Tengstedt, C., Fahlman, M., Robinson, N., & Berggren, M. A solid-state organic electronic wettability switch, Adv. Mater., 2004, 16, 316–320.

[8]   Whittell, G. R., & Manners, I. Metallopolymers: New multifunction materials, Adv. Mater., 2007, 19(21), 3439–3468.

[9]   Abbott, N.L., & Whitesides, G.M. Potential-dependent wetting of aqueous solutions on self-assembled monolayers formed from 15-(ferrocenylcarbonyl) pentadecanethiol on gold, Langmuir, 1994, 10(5), 1493–1497.

[10]  Min, U., & Chang, J.H. Thermotropic liquid crystalline polyazomethine nanocomposites via in situ interlayer polymerization, Mater. Chem. Phys., 2011, 129, 517–522.

[11]  Iwan, A., Palewicz, M., Chuchmała, A., Gorecki, L., Sikora, A., Mazurek, B., & Pasciak, G. Opto (electrical) properties of new aromatic polyazomethines with fluorine moieties in the main chain for polymeric photovoltaic devices, Synth. Met., 2012, 162, 143–153.

[12]  Senel, M., Evik, E.C., & Abasiyanik, M.F. Amperometric hydrogen peroxide biosensor based on covalent immobilization of horseradish peroxidase on ferrocene containing polymeric mediator, Sens. Actuators B Chem., 2010, 145(1), 444–450.

[13]  Soldatkin, A.P., Montoriol, J., Sant, W., Martelet, C., & Jaffrezic-Renault, N. N. Mat. Sci. Eng. C, 2002, 21, 75. Development of potentiometric creatinine-sensitive biosensor based on ISFET and creatinine deiminase immobilised in PVA/SbQ photopolymeric membrane Materials Science and Engineering: C, 21, 1–2, 1 September 2002, 75–79

[14]  Silvana, A., Barthelmebs, L., & Marty, J.L. Immobilization of acetylcholinesterase on screen-printed electrodes: Comparative study between three immobilization methods and applications to the detection of organophosphorus insecticides, Anal. Chim. Acta., 2002, 464, 171–180.

[15]  Dang, T.T.M., Park, S.J., Park, J.W., Chung, D.S., Park, C.E., Kim, Y.H., & Kwon, S.K. Synthesis and characterization of poly(benzodithiophene) derivative for organic thin film transistors, J. Polym. Sci. Part A: Polym. Chem., 2007, 45, 5277–5284.

[16]  Kim, B., Miyamoto, Y., Ma, B., & Fréchet, J. Photocrosslinkable polythiophenes for efficient, thermally stable, organic photovoltaics, Adv. Funct. Mater., 2009, 19, 2273–2281.

[17]  Contoret, A., Farrar, S., Oneill, M., Nicholls, J., Richards, G., Kelly, S., & Hall, A. The photopolymerization and cross-linking of electroluminescent liquid crystals containing methacrylate and diene photopolymerizable end groups for multilayer organic light-emitting diodes, Chem. Mater., 2002, 14(4), 1477–1487.

[18]  Charpoy, L.L. Recent advances in liquid crystalline polymers, Chapoy, L. L. (Ed.). Elsevier Appl. Sci., London, 1985, 89–95.

[19]  Ravikrishnan, A., Sudhakara, P., & Kannan, P. Liquid crystalline and photoactive poly[4,4′-stilbeneoxy]alkylbiphenylphosphates Poly. Degrad. Stab., 2008, 93, 1564–1570.

[20]  David, R., Angels, S., & Ana, M. New dimeric LC-epoxyimine monomers with oxyethylene central spacers: Crosslinking study, Polymer., 2003, 44, 2621–2629.

[21]  Ravikrishnan, A., Sudhakara, P., & Kannan, P. Liquid crystalline and photocrosslinkable poly (4,4′-stilbeneoxy) alkylarylphosphates, Poly. Eng. Sci., 2012, 52(3), 598–606.

[22]  Palomera, N., Vera, J.L., Melendez, E., Ramirez-Vick, J.E., Tomar, M.S., Arya, S.K., & Singh, S.P. Redox active poly(pyrrole-N-ferrocene-pyrrole) copolymer based mediator-less biosensors, J. Electroanal. Chem., 2011, 658, 33–37.

[23]  Soon, G.H., Deasy, M., Worsfold, O., & Dempsey, E. Synthesis, co-polymerization, and electrochemical evaluation of novel ferrocene-pyrrole derivatives, Anal. Lett., 2011, 44, 1976–1995.

[24] Senal, M. Construction of reagentless glucose biosensor based on ferrocene conjugated polypyrrole, Synth. Met, 2011, 161, 1861–1868.

[25] Yang, W., Zhou, H., & Sun, C. Synthesis of ferrocene-branched chitosain derivatives: Redox polysaccharides and their application to reagentless enzyme-based biosensors, Macromol., Rapid Commun, 2007, 28, 265–270.

[26] Nagarale, R.K., Lee, J.M., & Shin, W. Electrochemical properties of ferrocene modified polysiloxane/chitosan nanocomposite and its application to glucose sensor, Electrochim. Acta, 2009, 54, 6508–6514.

[27] Zheng, L., Xiong, L., Li, J., Li, X., Sun, J., Yang, S., & Xia, J. Synthesis of a novel b-cyclodextrin derivative with high solubility and the electrochemical properties of ferrocene-carbonyl-b-cyclodextrin inclusion complex as an electron transfer mediator, Electrochem. Commun, 2008, 10, 340–345.

[28] Zheng, L,., Li, J., Xu, J., Xiong, L., Zheng, D., Liu, Q., Liu, W., Li, Yang,, S., & Xia, J. Improvement of amperometric glucose biosensor by the immobilization of FcCD inclusive complex and carbon nanotube, Analyst, 2010, 135, 1339–1344.

[29] Choi, S.J., Choi, B.G., & Park, S.M. Electrochemical sensor for electrochemically inactive beta-D(+)-glucose using alpha-cyclodextrin template molecules, Anal., Chem, 2002, 74, 1998–2002.

[30] Ng, S.C., Chan, H.S.O., Wong, P.M.L., Tan, K.L., & Tan, B.T.G. Novel heteroarylene polyazomethines: Their synthesis and characterizations, Polymer., 1998, 39(20), 4963–4968.

[31] Trikalitis, P.N., Petkov, V., & Kanatzidis, M.G. Structure of redox intercalated $(NH_4)_{0.5}V_2O_5,mH_2O$ xerogel using the pair distribution function technique, Chem. Mater., 2003, 15, 3337–3342.

[32] Li, H., Chi, W., Liu, Y., Yuan, W., Li, Y., Li, Y., & Tang, B.Z. Metal-free poly-cycloaddition of activated azide and alkynes toward multifunctional polytriazoles: Aggregation-induced emission, explosive detection, fluorescent patterning, and light refraction, Macromol. Rapid. Commun., 2017, 38(18), 1700070 (7)

[33] Miller, J.S., Zhang, J.H., & Reiff, W.M. Structural and [57]Fe Mossbauer characterization of 1: 1 and 2:3 ferrocenium salts of 7,7,8,8-tetracyanoperfluoro-p-quinodimethane, Inorg. Chem., 1987, 26(4), 600–608.

[34] Okuno, S., & Matsubayashi, G. Intercalation of ferrocene and (ferrocenylalkyl) ammonium halides into the geL-V25 interlayer space, Bull. Chem. Soc. Jpn., 1993, 66(2), 459–463.

[35] Wang, P.H., Ho, M.S., Yang, S.H., Chen, K.B., & Hsu, C.S. Synthesis of thermal-stable and photo-crosslinkable polyfluorenes for the applications of polymer light-emitting diodes, J. Polym. Sci A Polym Chem, 2010, 48, 516–524.

[36] Wolfe, J.P., Rennels, R.A., & Buchwald, S.L. Intramolecular palladium-catalyzed aryl amination and aryl amidation, Tetrahedron, 1996, 52(21), 7525–7546.

[37] Bellmann, E., Shaheen, S.E., Grubs, R.H., Marder, S.R., Kippelen, B., & Peyghambarian, N. Organic two-layer light-emitting diodes based on high-$t_g$ hole-transporting polymers with different redox potentials, Chem. Mater., 1999, 11(2), 399–407.

[38] Gangadhara, Kishore K. Novel photocrosslinkable liquid-crystalline polymers: Poly[bis (benzylidene)] esters, Macromolecules, 1993, 26(12), 2995.

[39] Deepa, G., Balamurugan, R., & Kannan, P. Photoactive liquid crystalline polyesters based on bisbenzylidene and pyridine moieties, J. Mol. Struct., 2010, 963(2), 219–227.

[40] Sakthivel, P., & Kannan, P. Novel thermotropic liquid crystalline-cum-photocrosslinkable polyvanillylidene alkyl/arylphosphate esters, J. Polym. Sci. A Polym. Chem., 2004, 20(4), 5215–5226.

[41] Petersen, J.M. In: Price CC ed. Organic synthesis, Wiley, New York, 1953, Vol. 53.

[42] Craig, A.A., & Imrie, C.T. Effect of spacer length on the thermal properties of side-chain liquid crystal polymethacrylates. 2. Synthesis and characterization of the poly[.omega.-(4'-cyanobiphenyl-4-yloxy)alkyl methacrylate]s, Macromolecules, 1995, 28(10), 3617–3624.

[43] Gonsalves, K., Zhan-ru, L., & Rausch, M.D. Ferrocene-Containing Polyamides and Polyureas, J. Am. Chem. Soc, 1984, 106, 3862–3863.

[44] Li, G., Liu, Q., Liao, B., Chen, L., Zhou, H., Zhou, Z., Xia, B., Huang, J., & Liu, B. Synthesis of novel ferrocene-based conjugated microporous polymers with intrinsic magnetism, Eur. Polym. J., 2017, 93, 556–560.

[45] Musgrave, R.A., Russell, A.D., Hayward, D.W., Whittell, G.R., Lawrence, P.G., Gates, P.J., Green, J.C., & Manners, I. Main-chain metallopolymers at the static–Dynamic boundary based on nickelocene, Nat. Chem., 2017, 9, 743–750.

[46] Dervisevic, M., Dervisevic, E., Senel, M., Cevik, E., Yildiz, H.B., & Camurlu, P. Construction of ferrocene modified conducting polymer-based amperometric urea biosensor, Enzyme Microb. Technol., 2017, 102, 53–59.

[47] Duan, W., Yin, X., Luo, C., Kong, J., Ye, F., & Pan, H. Microwave-absorption properties of SiOC ceramics derived from novel hyperbranched ferrocene-containing polysiloxane, J. Eur. Ceram. Soc., 2017, 37, 2021–2030.

[48] Paquette, J.A., Kenaree, A.R., & Gilroy, J.B. Metal-containing polymers bearing pendant nickel(II) complexes of Goedken's macrocycle, Polym. Chem., 2017, 8(14), 2164–2172.

[49] Zhang, X., Shen, Y., Zhang, Y., Shen, G., Xiang, H., & Long, H. A label-free electrochemical immunosensor based on a new polymer containing aldehyde and ferrocene groups, Talanta, 2017, 164, 483–489.

[50] Fatima, K., Gul, A., Akhter, Z., & Rahman, R. Ferrocene based azo chromophore functionalized polyesters and its organic analogue: Synthesis, structural elucidation, and thermal behavior and IV characterization, J. Inorg. Organomet. Polym. Mater., 2017, 27(2), 474–480.

[51] Patil, Y., Misra, R., Singhal, R., & Sharma, G.D. Ferrocene-diketopyrrolopyrrole based non-fullerene acceptors for bulk heterojunction polymer solar cells, J. Mater. Chem. A, 2017, 5, 13625–13633.

[52] Folkertsma, L., Zhang, K., Czakkel, O., De Boer, H.L., Hempenius, M.A., Van Dan Berg, A., Odijk, M., & Vancso, G.J. Synchrotron SAXS and impedance spectroscopy unveil nanostructure variations in redox-responsive porous membranes from poly(ferrocenylsilane) poly(ionic liquid)s, Macromolecules, 2017, 50(1), 296–302.

[53] Gan, L., Song, J., Guo, S., Janczewski, D., & Nijhuis, C.A. Side chain effects in the packing structure and stiffness of redox-responsive ferrocene-containing polymer brushes, Eur. Polym. J., 2016, 83, 517–528.

[54] Bizid, S., Mlika, R., Haj Said, A., Chemli, M., & Korri-Youssoufi, H. Functionalization of MWCNTs with Ferrocene-poly(p-phenylene) and effect on electrochemical properties: Application as a sensing platform, Electroanalysis, 2016, 28, 2533–2542.

[55] Homann-Muller, T., Rieger, E., Alkan, A., & Wurm, F.R. N-Ferrocenylsulfonyl-2-methylaziridine: The first ferrocene monomer for the anionic (co)polymerization of aziridines, Polym. Chem., 2016, 7, 5501–5506.

[56] Shekurov, R., Miluykov, V., Kataeva, O., Krivolapov, D., Sinyashin, O., Gerasimova, T., Katsvuba, S., Kovalenko, V., Krupskaya, Y., Kataev, V., Buchner, B., Senkovska, I., & Kaskel, S. Reversible water-induced structural and magnetic transformations and selective water adsorption properties of Poly(manganese 1,1′-ferrocenediyl-bis(H-phosphinate), Cryst. Growth. Des., 2016, 16(9), 5084–5090.

[57] Xiao, F., Yue, L., Li, S., & Li, X. Conjugation of cytochrome c with ferrocene-terminated hyperbranched polymer and its influence on protein structure, conformation and function, Spectrochim. Acta. A Mol. Biomol. Spectrosc., 2016, 162, 69–74.

[58] Maslakci, N.N., Eren, E., Topel, S.D., Cin, G.T., & Oksuz, AU. Electrospun plasma-modified chitosan/poly(ethylene terephthalate)/ferrocenyl-substituted N-acetyl-2-pyrazoline fibers for phosphate anion sensing, J. Appl. Polym. Sci., 2016, 133(17), 43344 (1–7).

[59] Bizid, S., Mlika, R., Haj Said, A., Chemli, M., & Korri Youssoufi, H. Investigations of poly(p-phenylene) modified with ferrocene and their application in electrochemical DNA sensing, Sens. Actuators B Chem., 2016, 226, 370–380.

[60] Xiang, J., Sato, K., Tokue, H., Oyaizu, K., Ho, C.L., Nishide, H., Wong, W.Y., & Wei, M. Synthesis and charge-discharge properties of organometallic copolymers of ferrocene and triphenylamine as cathode active materials for organic-battery applications, Eur. J. Inorg. Chem., 2016, 2016, 1030–1035.

[61] Xiang, J., Wang, T.K., Zhao, O., Huang, W., Ho, C.L., & Wong, W.Y. Ferrocene-containing poly (fluorenyl ethynylene)s for nonvolatile resistive memory devices, J. Mater. Chem., 2016, 4(5), 921–928.

[62] Riquelme, J., Garzon, C., Bergmann, C., Geshev, J., & Ouijada, R. Development of multifunctional polymer nanocomposites with carbon-based hybrid nanostructures synthesized from ferrocene, Eur. Polym. J., 2016, 75, 200–209.

[63] Khalid, H., Wang, L., Yu, H., Akram, M., Abbasi, N.M., Sun, R., Saleem, M., Zain-Ul-abdin, R., & Chen, Y. Synthesis of soluble ferrocene-based polythiophenes and their properties, J. Inorg. Organomet. Polym. Mater., 2015, 25(6), 1511–1520.

[64] Guo, W., & Lei, Z. Redox-responsive supramolecular polymer based on β-cyclodextrin and the ferrocene-decorated main chain of PAA, J. Mater. Res., 2015, 30(21), 3201–3210.

[65] Nguema, R.W., Dzang, E., Lejars, M., Brisset, H., Raimundoand, J.M., & Bressy, C. RAFT-synthesized polymers based on new ferrocenyl methacrylates and electrochemical properties, RSC. Adv., 2015, 5(94), 77019–77026.

[66] Deng, F., Ding, W., Peng, Z., Li, L., Wang, X., Wan, X., Cheng, L., & Li, M. RETRACTED: The synthesis of ferrocene-based mixed-metal coordination polymer microspheres and their application in hydrogen storage, J. Alloys. Compd., 2015, 647, 1111–1120.

[67] Xiang, J., Burges, R., Haupler, B., Wild, A., Schubert, U.S., Ho, C.L., & Won, W.Y. Synthesis, characterization and charge-discharge studies of ferrocene-containing poly (fluorenylethynylene) derivatives as organic cathode materials, Polymer, 2015, 68, 328–334.

[68] Deng, Z., Yu, H., Wang, L., & Zhai, X. A novel ferrocene-containing polymer based dispersant for noncovalent dispersion of multi-walled carbon nanotubes in chloroform, J. Organomet. Chem., 2015, 791, 274–278.

[69] Gul, A., Sarfraz, S., Akhter, Z., Siddiq, M., Kalsoom, S., Perveen, F., Ansari, F.L., & Mirza, B. Synthesis, characterization and biological evaluation of ferrocene based poly(azomethene) esters, J. Organomet. Chem., 2015, 779, 91–99.

[70] Liu, L., Rui, L., Gao, Y., & Zhang, W. Self-assembly and disassembly of a redox-responsive ferrocene-containing amphiphilic block copolymer for controlled release, Polym. Chem., 2015, 6(10), 1817–1829.

[71] Erden, P.E., Kacar, C., Ozturk, F., & Kilic, E. Amperometric uric acid biosensor based on poly (vinylferrocene)-gelatin-carboxylated multiwalled carbon nanotube modified glassy carbon electrode, Talanta, 2015, 134, 488–495.

[72] Ayranci, R., Demirkol, D.O., Ak, M., & Timur, S. Ferrocene-functionalized 4-(2,5-Di(thiophen-2-yl)-1H-pyrrol-1-yl)aniline: A novel design in conducting polymer-based electrochemical biosensors, Sensors, 2015, 15(1), 1389–1403.

[73] Liu, Q., Tang, Z., Wu, M., Liao, B., Zhou, H., Ou, B., Yu, G., Zhou, Z., & Li, X. Novel ferrocene-based nanoporous organic polymers for clean energy application, RSC. Adv., 2015, 5(12), 8933–8937.

[74] Su, C., Ji, L., Xu, L., Zhu, X., He, H., Lu, Y., Ouyang, M., & Zhang, C. A novel ferrocene-containing aniline copolymer: Its synthesis and electrochemical performance, RSC. Adv., 2015, 5(18), 14053–14060.

[75]  Goswami, T.K., Gadadhar, S., Balaji, B., Gole, B., Karande, A.A., & Chakravarty, A.R. Ferrocenyl-L-amino acid copper(II) complexes showing remarkable photo-induced anticancer activity in visible light, Dalton. Trans., 2014, 43(31), 11988–11999.

[76]  Zhong, H., Wang, G., Song, Z., Li, X., Tang, H., Zhou, Y., & Zhan, H. Organometallic polymer material for energy storage, Chem. Comm., 2014, 50(51), 6768–6770.

[77]  H. Zhong, G. Wang, Z. Song, X. Li, H. Tang, Y. Zhoua, H. Zhan, Organometallic polymer material for energy storage Chem. Commun., 2014, 50, 6768-6770

[78]  Sun, Y., Yu, H., Wang, L., Zhao, Y., Ding, W., Ji, J., Chen, Y., Ren, F., Tian, Z., Huang, L., Ren, P., & Tong, R. Preparation of ferrocene-based coordination polymer microspheres and their application in hydrogen storage, J. Inorg. Organomet. Polym., 2014, 24, 491–500.

[79]  Drelich, J., Chibowski, E., Meng, D.D., & Terpilowski, K. Hydrophilic, and superhydrophilic surfaces and materials, Soft Matter, 2011, 7, 9804–9828.

# 6 Transition metal-based coordination polymers

**Abstract:** Inorganic coordination polymer has a demand in the material science research world, supplying a distinct method of synthesis, which is developed from various molecular blocks along with the different correlation between them. This chapter gives the outline of future generation coordination polymers and its structural arrangement such as porous, nonporous, and nanospace polymer composites.

**Keywords:** porous structures, nonporous structures, nanoscale transition metal-polymer, synthesis

## 6.1 Introduction

Nowadays, evolution in the synthesis of polymer materials coordinated with transition metal has the center of attraction in the research world. The great variety of structures and its properties of the polymers, as well as facile synthesis, has the center of attractions for many research scientists. Transition metal coordination polymers have great unique properties in various fields such as inorganic and organic chemistry, material chemistry, electrochemistry, biochemistry, pharmacology, and electroanalytical chemistry along with a special variety of applications [1–4].

Coordination polymers can be segregated into two types as per their structure and composition into nonporous coordination polymers and porous coordination polymers. Generally, porous coordination polymers are more flexible along with high surface area. Coordination polymers can differentiate as per the structural pattern such as one-dimensional (1D), two-dimensional (2D), and three-dimensional (3D). Figure 6.1 represents 1D coordination polymeric structure, in a consecutive way along $x$-axis. Two-dimensional coordination polymeric structure go on toward $x$- and $y$-axes. Three-dimensional coordination polymeric structure move toward $x$-, $y$-, and $z$-axes (three directions). As van der Waal's forces as well as hydrogen bonding ($\pi$–$\pi$ interactions) disturb dimensionalities of coordination polymers, it affects the structure. The dimensionality of coordination polymers was possibly shown to be 2–10 coordination numbers.

This number of atoms surrounding a central atom in a complex is called coordination number. Coordination number and geometry can be estimated by the number of systematic distributions and also depends upon the size of cations. Transition metal atom like copper has completely filled d-orbitals in their outer shells due to their electronic orbital, and as a result, shows a variety of coordination geometries. Frequently used polymer ligands are polypyridines [5], polypyrrole [6], phenanthrolines [7], polyaniline [8], polycarboxylates [9], and hydroxyquinolines [10] in coordination polymers. These polymers containing oxygen, nitrogen, sulfur, and phosphorus atoms are commonly used to coordinate a covalent bond or as a binding site. As a result, the

https://doi.org/10.1515/9781501514609-007

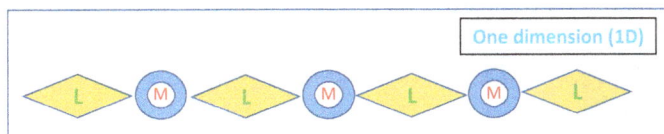

**Figure 6.1:** Schematic representation of one-dimensional (1D) structure coordination polymers.

formation of coordinate covalent bond between metal and ligand forms complex as per the hard–soft acid–base theory. According to the contradictory study, small-scale nonpolarized hard ligands coordinate with nonpolarized hard metals and on the other hand highly polarized soft metals will correlate toward the highly polarized soft ligands. An additional aspect about the structure of coordination polymers depends on the style of ligand as either rigid or flexible. Flexible ligand shows various combinations in the structure due to its orientation such as rotating, bending style. Figure 6.2 shows the same ligand structure, but with different orientation, and Figure 6.3 shows various structures based on their flexibility to ligand. A rigid ligand has the inability to rotate within a structure.

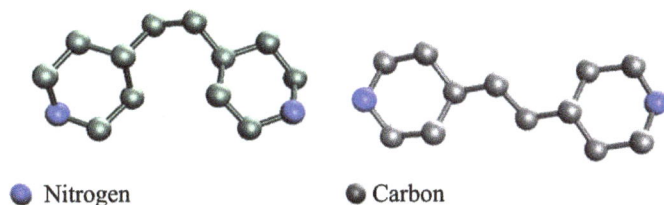

● Nitrogen          ● Carbon

**Figure 6.2:** Different orientations (gauche and anti) of ligand.

The structural factors affecting coordination polymers are a counter ion, pH, temperature, and crystallization [11, 12].

## 6.2  Structure of coordination polymers

Transition metal coordination polymer synthesis can be classified into three ways: porous coordination polymers, nonporous coordination polymers, and transition metal coordination polymers. Coordination polymer synthesis depends on two central components like linkers and connectors. The vital role of connectors and linkers is the coordination number and their binding sites. Transition metals are exploited as multiskilling connectors in the synthesis of coordination polymers.

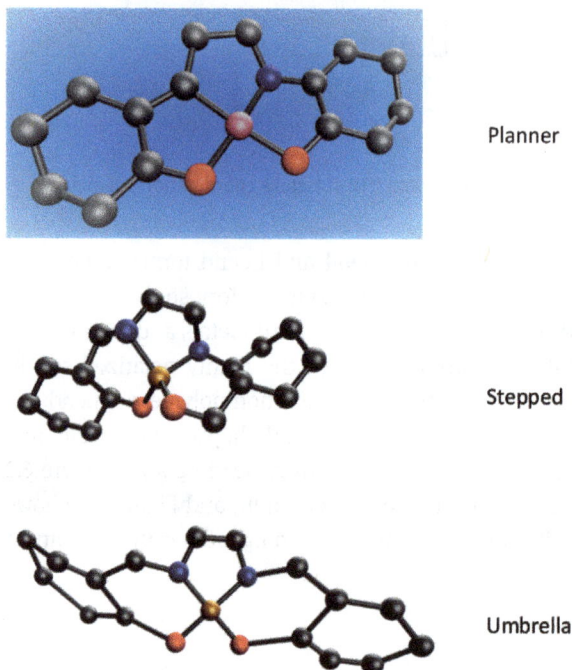

Planner

Stepped

Umbrella

**Figure 6.3:** Three different molecular conformations (nitrogen: blue; oxygen: red; carbon: gray; and metal: golden) of salen complexes.

## 6.2.1 Porous structures

### 6.2.1.1 Dots (0D)
Coordination polymer solids are classified into two categories, solid with windows and solid without windows. The size of these windows is comparatively tinier than guest molecules.

The percolated 3D network structure of $\{Zn(CN)(NO_3)(tpt)_{2/3}\}.3/4\ C_2H_2Cl_4.3/4CH_3OH\}_n$ gives an impassable boundary to tiny molecules, effectively isolating cavity.

### 6.2.1.2 Channels (1D space)
The coordination polymer of Cu(II), $[Cu_2(pzdc)_2\ (ps)]_n$ has showed pillared layer morphology and is applicable for the layout of porous structure (CPL-1; coordination polymer-1). The central metal copper (II) developed distorted square pyramidal coordination by using three carboxylate oxygen atoms, one nitrogen atom of pyz, and one nitrogen atom of pzdc. The pillared layer morphological structure developed from Cu (II) and pzdc is connected by pyz ligands [13].

### 6.2.1.3 Layers (2D space)

Researcher coined new class in a 2D network structure by using two different metals represented as $[M^{II} M^{III} (ox)_3]_n$ or of bimetallic phase 2D network structure. Metals such as $M^{II}$ = Fe, Co, Cu, Zn, Mn, and $M^{III}$ = Fe, Cr are employed in the 2D network of coordination polymers. Okawa et al. reported for the first time this concept by using same transition metals such as ferri and ferro or antiferromagnets at critical temperatures between 5 and 45 K. It forms an extended anionic network structure due to oxalate-bridged hexagonal strips of the two metal atoms. The organometallic decamethylmetallocenium cations are $[Y^{III} (Cp\star)2]$ $(Y^{III}$ = Fe, Co; $Cp\star$ = $C_3Me_5$) [14, 15].

### 6.2.1.4 Layers (3D space)

A 3D type of chiral framework structure formed for $Zn(mpda)_{0.5}bix. (H_2O)_{1.5}]_n$. The Zn(II) atom adopts a distorted square-pyramidal coordination symmetry. Chiral sheet format structure observed in the ab plane (parallel type), and Zn(II) ions are connected to the carboxylate group along with syn-/anti-conformation. Zn atoms and the ligands are connected in the form of metal at the center, representing the framework style [16].

### 6.2.2 Nonporous

The nonporous coordination polymer gel developed by Xu and coworkers, in this general combination of four bonds such as hydrogen bonds, covalent bond, ionic bonds, and at the last hydrophobic interaction, create a gel-type material. Figure 6.4 (a) shows a good coordination in between transition metal (Pd) and tetradentate ligands, which produced a more stable metallogel.

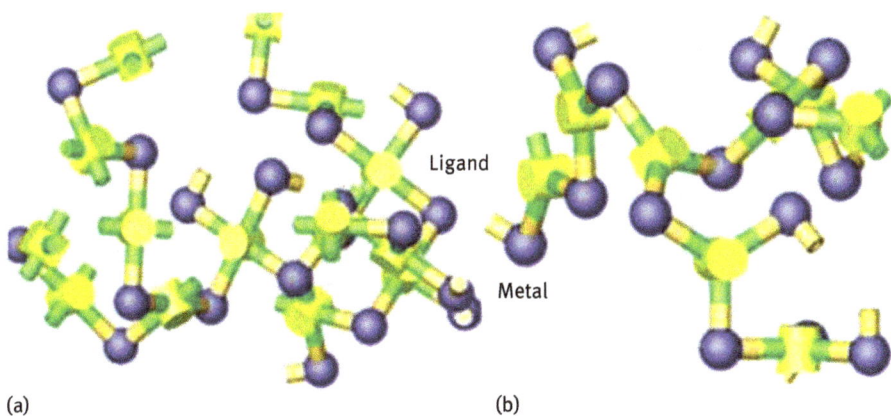

Ligand

Metal

(a)　　　　　　　　　　　　　　　　　　(b)

**Figure 6.4:** Teradentate ligand (a) and tridentate ligand (b) to correlate with a transition metal ion (Pd) to form a 3D network of coordination polymers.

Cross-linked 3D structured metallogel (nonporous coordination polymer) materials were developed by using multidentate ligands or chelates and transition metals that have extra coordination sites.

The porphyrin-based tetradentate ligand/chelates have a planar and rigid-type structure, which is shown in Figure 6.5(a). Combining reactions 6.5(a) and 6.5(b) in the presence of dimethylsulfoxide formed the gel-type structure, after 70 days [17].

(a)                                    (b)

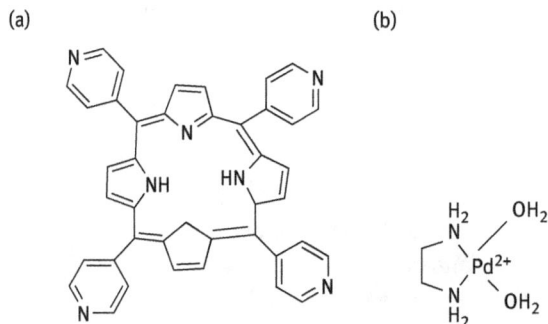

**Figure 6.5:** (a) Porphyrin-based tetradentate ligand/chelate and (b) structure of $[Pd(en)(H_2O)_2]^{2-}$ (en = ethylenediamine).

### 6.2.3 Nanosized transition metal coordination polymers

Nanoscale transition metal coordinates with polymers have the center of attraction in the research world, because of their facile synthesis, bulk manufacturing, and significantly lighter in weight than other metals. Nanocomposites are nanoscale materials encapsulated in a polymer matrix. In the case of nanoparticles, the particle morphology and proportion play a vital role. The electrical, mechanical, thermal, and optical properties of the polymer nanocomposites vary.

## 6.3 Synthetic methods

Several ways are adopted for the preparation of polymer–metal nanocomposites. Most of the nanocomposites are synthesized by chemical oxidative polymerization. Chemical oxidative polymerization can be classified into ex situ and in situ methods [18, 19].

In the ex situ method, the nanoparticles are synthesized and embedded within the polymer, before the preparation of the conjugated polymer. The conjugated polymers are synthesized first by oxidative polymerization (chemical or electrochemical) of the corresponding monomer. This is then followed by the addition of nanoparticles into the polymer matrix chemically/electrochemically or physically. The preparation of organic polymer and metal nanoparticles is achieved independently. Polymer matrix and metal nanoparticles are mechanically or physically blended to form polymer–metal nanocomposites. A drawback of ex situ method is that the metal nanoparticles are more attracted toward monomer active sites than the polymer active sites, thus rendering the chemical bonding of polymer with nanoparticles passive. The in situ technique, on the other hand, is employed nowadays for the synthesis of polymer–metal nanocomposites. This is based on the reduction of metal ions scattered in polymer matrices and is a facile and an effective route. This method permits a single step of nanocomposite preparation, in which diffusion of the nanoparticles is more, which improves the properties of the nanocomposites.

Researchers have found various methods for the synthesis of polymer-based transition metal nanoparticle composites such as melt intercalation [20, 21], template synthesis [22, 23], and the sol–gel process [24–26].

Some examples of the synthesis of nanocomposites are the synthesis of polyaniline-$TiO_2$ using chemical oxidation polymerization. This was carried out by blending aniline with hydrophobic $ILPF_6$ (ionic liquid), $n$-butanol, and nitric acid in the same reaction mixture. The mixture is stirred, and dispersed $TiO_2$ nanoparticles are added to it [27]. Ammonium persulfate is added for the initiation of the reaction. Preparation of polyaniline–Cu nanocomposite is obtained by blending aniline with HCl solution and adding colloidal copper nanoparticles followed by stirring. Further, ammonium peroxydisulfate is added for initiation of polymerization reactions, after 12 h reaction completed [28]. Brooks and his coworkers have reported the facile synthesis of polyaniline–Au nanocomposite, where gold chloride solution is blended with anilinium hydrochloride [29].

A concise survey showed that there are two main kinds of nanoscale composites of polymers with transition metals: (1) transition metal nanoparticles encapsulated into the polymer matrix and (2) metal core nanoparticles concealed with a polymer shell. The advantages of incorporation of nanoscale transition metals into polymer matrices find application as catalysts, optical and dental application, in microelectronics, in abrasion-resistant coatings, cathode materials for rechargeable lithium batteries, and sensors [30, 31].

ZnO nanoparticles have been synthesized by zinc acetate. Tentative reaction mechanism of poly($N$-ethyl aniline)/transition metal nanocomposites is shown in Figure 6.6. The synthesized nanoparticles (Zn) are mixed with the monomer ($N$-ethyl aniline). Ammonium persulfate is used as an initiated oxidizing agent for the polymerization reaction. Poly($N$-ethyl aniline) is encapsulated to Zn metal nanoparticles as shown in Figure 6.7 [32].

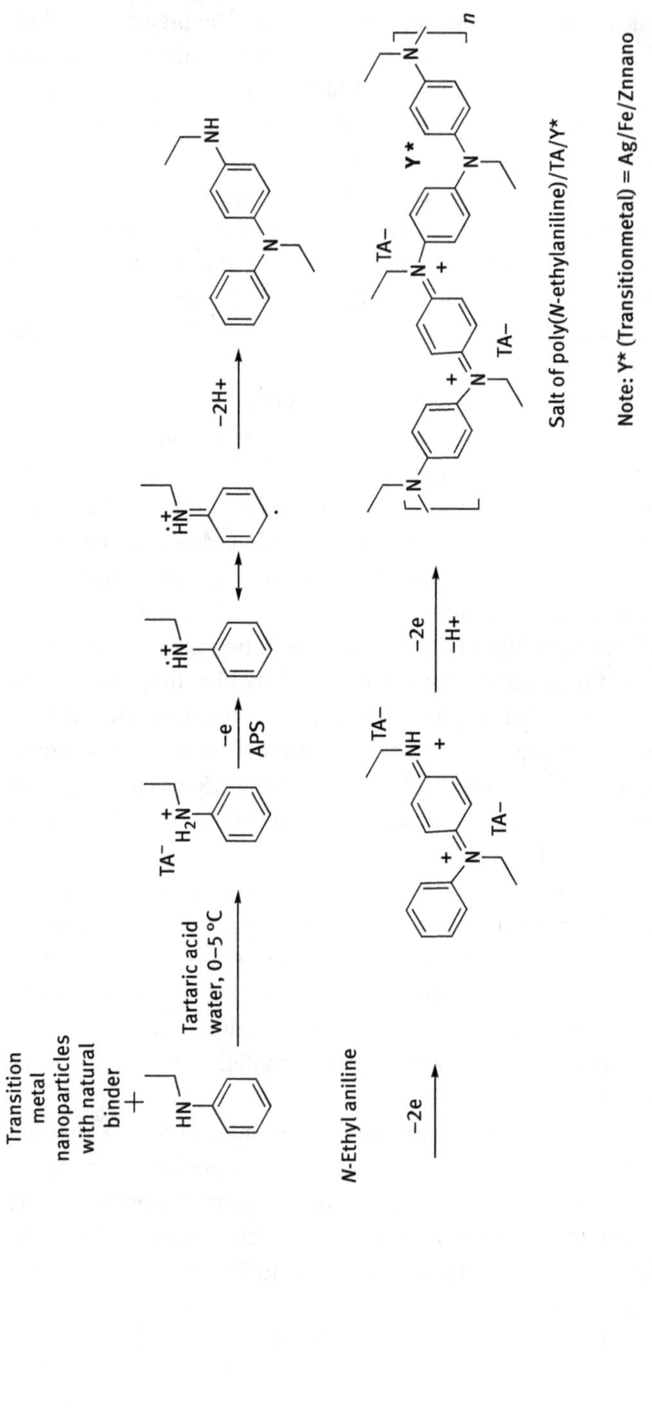

**Figure 6.6:** Tentative mechanism of formation of poly(N-ethyl aniline)/Zn nanocomposite.

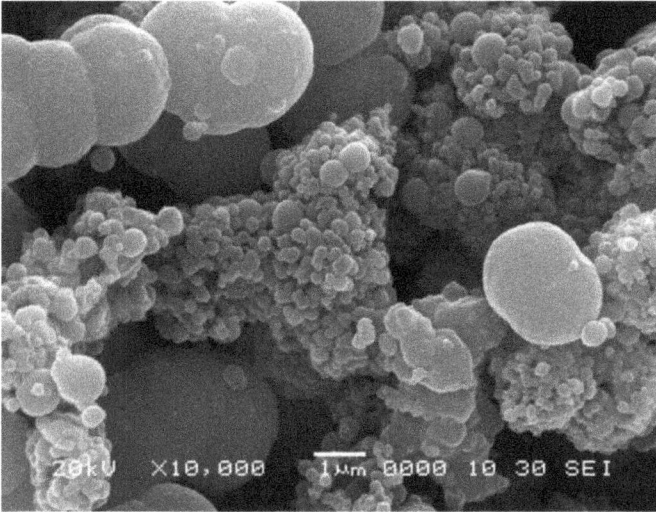

**Figure 6.7:** Coordination of poly (*N*-ethyl aniline) along with Zn.

## 6.4 Properties

In mechanical properties, the variation in toughness and modulus fully depends on the degree of interaction between the polymer and the nanoparticles. The electrical and optical properties of nanocomposites, on the other hand, can be enhanced by adding nanoparticles. Nanocomposite morphology and size can lead to changes in their electrical and optical properties. Nanocomposites possess conductive or optical properties and catalytic activity for organic reaction conversions.

## 6.5 Applications

The coordination, transition metal, polymer composites are especially used as catalysts in organic reactions and are also used in detection of hazardous gases such as ammonia, $H_2S$, and humidity [33–37].

## 6.6 Conclusion

Transition metal coordinates with polymeric materials expressed unique vista in the research world. The combination of both materials to retain their exclusive structural features in the form of porous and nonporous fashion. Cost-effective nanoscale transition metal nanoparticles with polymers created the boom in material science-based applications due to their special structural features and unique properties.

# References

[1] Chaudhary, R.G., Tanna, J.A., Gandhare, N.V., Bagade, M.B., Bhuyar, S.S., Gharpure, M.P., & Juneja, H.D. Synthesis and characterization of metal coordination polymers with fumaroylbis (paramethoxyphenylcarbamide) by using FTIR, XRD, SEM and TG techniques, J. Chinese Adv. Mater. Soc., 2015, 3(3), 177–187.

[2] Bureekaew, S., Shimomura, S., & Kitagawa, S. Chemistry and application of flexible porous coordination polymers, Sci. Technol. Adv. Mater., 2008, 9, 014108 (12pp).

[3] Leong, W. L., & Vittal, J. J. One-dimensional coordination polymers: complexity and diversity in structures, properties, and applications, Chem. Rev., 2011, 111(2), 688–764.

[4] Foo, M.L., Matsuda, R., & Kitagawa, S. Functional hybrid porous coordination polymers, Chem. Mater., 2014, 26(1), 310–322.

[5] Juris, A., Balzani, V., Barigelletti, F., Campagna, S., Belser, Pl., & Von Zelewsky, A. Ru (II) polypyridine complexes: photophysics, photochemistry, electrochemistry, and chemiluminescence}, Coord. Chem. Rev., 1998, 84, 85–277.

[6] Alain, D., & Jean-Claude, M. Polypyrrole films containing metal complexes: syntheses and applications, Coord. Chem. Rev., 1996, 147, 339–371.

[7] Yanjuan, Q., Yonghui, W., Changwen, H., Minhua, C., Li, M., & Enbo, W. A new type of single-helix coordination polymer with mixed ligands [M2 (phen) 2 (e, a-cis-1, 4-chdc) 2 (H2O) 2] n (M= Co and Ni; phen= 1, 10-phenanthroline; chdc= cyclohexanedicarboxylate)}, Inorg. chem., 2003, 42(25), 8519–8523.

[8] Tang, L., Wu, T., & Kan, J. Synthesis and properties of polyaniline--cobalt coordination polymer, Synth. Met., 2009, 159(15–16), 1644–1648.

[9] Elahi, S.M., Chand, S., Deng, W.H., Pal, A., & Das, M.C. Polycarboxylate-templated coordination polymers: role of templates for superprotonic conductivities of up to $10^{-1}$ S cm$^{-1}$, Angew. Chem. Int. Ed. Engl., 2018, 57(22), 6662–6666.

[10] Chen, H.X., Zhou, F., Ma, Y., Xu, X.P., Ge, J.F., Zhang, Y., Xu, Q.F., & Lu, J.M. Preparation of coordination polymers with 8-hydroxyquinoline azo benzenesulfonic acid as a planar multidentate ligand and the study of their photochemical and photo-stability properties, Dalton Trans., 2013, 42(14), 4831–4839.

[11] Pan, L., Frydel, T., Sander, M.B., Huang, X., & Li, J. The effect of pH on the dimensionality of coordination polymers, Inorg. Chem., 2001, 40(6), 1271–1283.

[12] Bureekaew, S., Shimomura, S., & Kitagawa, S. Chemistry and application of flexible porous coordination polymers, Sci. Technol. Adv. Mater., 2008, 9(1), 014108.

[13] Kondo, M., Okubo, T., Asami, A., Noro, S., Yoshitomi, T., Kitagawa, S., Ishii, T., Matsuzaka, H., & Seki, K. Rational synthesis of stable channel-like cavities with methane gas adsorption properties:[{Cu2 (pzdc)2 (L)n](pzdc= pyrazine-2, 3-dicarboxylate; L= a Pillar Ligand), Angew. Chem., 1999, 111(38), 140–143.

[14] Hamao, W., Motohiko, K., Okawa, T., Yuichi, K., Yoichiro, N., & Midori, G. Crystal and molecular structure of [i-Pr2Si] 4 and [(Me3SICH2) 2Si] 4 and some structural properties of cyclopolysilanes,[R1R2Si] n (n= 3–6), Appl. Organomet. Chem., 1987, 1(2), 157–169.

[15] Coronado, E., Galn-Mascars, J.R., Gmez-GarcVa, C.J., Ensling, J., & Gtlich, P. Hybrid molecular magnets obtained by insertion of decamethylmetallocenium cations into layered, bimetallic

oxalate complexes:[ZIIICp* 2][MIIMIII (ox) 3](ZIII= Co, Fe; MIII= Cr, Fe; MII= Mn, Fe, Co, Cu, Zn; ox= oxalate; Cp*= pentamethylcyclopentadienyl), Chem. Eur. J., 2000, 6, 552–563.

[16]   Ji, C., Zang, S.Q., Liu, J.Y., Li, J.B., & Hou, H.W. Assembly of 1,2,3,4-Benzenetetracarboxylic Acid and Zinc(II) Metal centers to a chiral 3D metal-organic framework: Syntheses, structure, and properties, Z. Naturforsch., 2011, 66b, 533–537.

[17]   Xing, B., Choi, M., & Xu, B. Design of coordination polymer gels as stable catalytic systems, Chem. A Eur. J., 2002, 8(21), 5028–5032.

[18]   Pande, N.S., Jaspal, D., & Ambekar, J. Poly (N-ethyl aniline)/Ag nanocomposite as humidity sensor, Int. J.Nanosci., 2017, 16(3), 1650037.

[19]   Segala, K., Dutra, R.L., Franco, C.V., Pereira, A.S., & Trindadeb, T.S. In situ and ex-situ preparations of ZnO/poly-{trans-[RuCl$_2$ (vpy) 4]/styrene} nanocomposites, J. Brazil. Chem. Soc., 2010, 21(10), 1986–1991.

[20]   Kawasumi, M., Hasegawa, N., Kato, M., Usuki, A., & Okada, A. Preparation and mechanical properties of polypropylene-clay hybrids, Macromolecules, 1997, 30(20), 6333–6338.

[21]   Vaia, R.A., & Giannelis, E.P. Lattice model of polymer melt intercalation in organically-modified layered silicates, Macromolecules, 1997, 30(25), 7990–7999.

[22]   Tomasko, D.L., Han, X., Liu, D., & Gao, W. Supercritical fluid applications in polymer nanocomposites, Curr. Opin. Solid St. M., 2003, 7(4), 407–412.

[23]   Carrado, K.A.;., & Xu, L. In situ synthesis of polymer-clay nanocomposites from silicate gels, Chem. Mater., 1998, 10(5), 1440–1445.

[24]   Mark, J.E. Some novel polymeric nanocomposites, Acc. Chem. Res., 2006, 39(12), 881–888.

[25]   Avadhani, C.V., & Chujo, Y. Polyimide–Silica gel hybrids containing metal salts: Preparation via the sol-gel reaction, Appl. Organomet. Chem., 1997, 11(2), 153–161.

[26]   Liu, J., Gao, Y., Wang, F., Li, D., & Xu, J. Preparation and characteristic of a new class of silica/polyimide nanocomposites, J. Mater. Sci., 2002, 37(14), 3085–3088.

[27]   Guo, Y., He, D., Xia, S., Xie, X., Gao, X., Zhang, Q., & Obare, S. Preparation of a novel nanocomposite of polyaniline core decorated with anatase-TiO$_2$ nanoparticles in ionic liquid/water microemulsion, J. Nanomater., 2011, 2012(906), 32–40.

[28]   Liu, A., Bac, LH., Kim, JS., Kim, BK., & Kim, JC. Synthesis and characterization of conducting polyaniline-copper composites, J.Nanosci. Nanotech., 2013, 13(11), 7728–7733.

[29]   Prakash, S., Rao, C.R., & Vijayan, M. Polyaniline–Polyelectrolyte–Gold (0) ternary nanocomposites: Synthesis and electrochemical properties, Electrochim. Acta, 2009, 54(24), 5919–5927.

[30]   Kang, Y.O., Choi, S.H., Gopalan, A., Lee, K.P., Kang, H.D., & Song, Y.S. Tuning of morphplogy of Ag nanoparticles in the Ag/polyaniline nanocomposites prepared by g-ray irradiation, J. Non-Cryst. Solid., 2005, 352, 463–468.

[31]   Gasaymeh, S.S., Radiman, S., Heng, L.Y., & Saion, E. Gamma irradiation synthesis and influence the optical and thermal properties of cadmium sulfide (CdS)/poly (vinyl pyrrolidone) nanocomposites, Am. J. Appl. Sci., 2010, 7(4), 500–508.

[32]   Pande, N.S., Jaspal, D., Warke, A., & Chabukswar, V.V. Eco-friendly synthesis, characterization of poly(n-ethyl aniline)/ZnO nanocomposite and its application as ammonia gas sensor, Polym. Res. J., 2016, 10(4), 239–248.

[33]   Liu, Chang-Feng., Moon, Doo-Kyung., Maruyama, T., & Yamamoto, T. Preparation of polymer blend colloids containing polyaniline or polypyrrole by Fe (II)-, Fe (III)-, and Cu (II)-H 2 O 2 catalyst system, Polym. J., 1993, 25(7), 775–779.

[34]   Pande, N.S., Jaspal, D., & Ambekar, J. Poly (N-ethyl aniline)/Ag nanocomposite as humidity sensor, Inter. J. Nanosci., 2017, 16(3), 1650037.

[35]   Pande, N., Jaspal, D., Malviya, A., & Warke, A. Synthesis, characterization, and application of poly (N-ethyl aniline)/iron nanocomposite, Inorg. Nano-Metal Chem., 2017, 47(7), 999–1003.

[36] Chauhan, N.P.S., Gholipourmalekabadi, M., & Mozafari, M. Fabrication of newly developed pectin −GeO2 nanocomposite using extreme biomimetics route and its antibacterial activities, J. Macromol. Sci. A, 2017, 54(10), 655–661.

[37] Batten, S.R., Hoskins, B. F., & Robson, R. Two interpenetrating 3D networks which generate spacious sealed-off compartments enclosing of the order of 20 solvent molecules in the structures of Zn (CN)(NO$_3$)(tpt) 2/3. cntdot. solv (tpt= 2, 4, 6-tri (4-pyridyl)-1, 3, 5-triazine, solv=. apprx. 3/4C$_2$H$_2$Cl$_4$. cntdot. 3/4CH$_3$OH or. apprx. 3/2CHCl$_3$. cntdot. 1/3CH$_3$OH), J. Am. Chem. Soc., 1995, 117, 5385–5386.

# 7 Geopolymers

**Abstract:** Geopolymers were synthesized by the following methods: stirring, grinding, and mixing. All geopolymers are structurally characterized by X-ray diffraction, scanning electron microscopy, $^{29}$Si magic angle spinning (MAS) nuclear magnetic resonance (NMR), and $^{2}$H MAS NMR spectroscopic methods. Geopolymers are used as adhesives and binders. Geopolymers are used for the removal of toxic and radioactive waste. They are used for making new cement for concrete.

**Keywords:** metakaolin, aluminum oxide ($Al_2O_3$), alkaline silicate solution, GFRP, POFA and sodium oxide ($Na_2O$)

## 7.1 Introduction

### 7.1.1 Geopolymers

Geopolymers are polymer framework structures. The alkali metal ion balances the charge of the tetrahedral aluminum, for example, obsidian. They are the rock-forming minerals and used as a raw material in the preparation of silicon-based polymers; therefore, they are known as geopolymer. In 1978, Joseph Davidovits coined the term geopolymer [1].

A geopolymer is a polymer that consists of repeating units and these repeating units are bound through geopolymerization. Geopolymerization is a process in which transfer of leached species occurs from the solid surface to the gel phase, which was followed by the condensation reaction of the gel phase to the solid binder.

Geopolymer was classified into groups as follows [2]
- pure inorganic and organic geopolymers and
- synthetic (man-made) geopolymers

Geopolymers are used as an alternative source in the preparation of structural concrete as compared to conventional Portland cement because their production is associated with much less $CO_2$ emissions [3]. Geopolymer possesses high compressive strength, high-temperature resistance, and stability under attack of acid as compared to Portland cement was reported by Duxson et al. [4]. The setting is a process in which material changes its state, that is, from the state of fluid to a solid state. Geopolymerization is a process, which includes the following nano-structural changes: dissolution, equilibration of dissolved species, gelation, reorganization, and polymerization.

https://doi.org/10.1515/9781501514609-008

The geopolymerization setting reactions involve the polycondensation of ortho-silicate ions [5, 6].

Li et al. investigated that the geopolymeric adsorbent was used for the removal of methylene blue dye [7]. Wang et al. studied the preparation of geopolymer for the removal of $Cu^{2+}$ ion from wastewater [8].

## 7.2 Synthetic methods

### 7.2.1 Metakaolin geopolymer – its evolution and setting behavior

#### 7.2.1.1 Materials required
Metakaolin, sodium hydroxide, silica powder, sodium oxide ($Na_2O$), aluminum oxide ($Al_2O_3$), silicon oxide ($SiO_2$), water ($H_2O$), and aluminum (Al) were used.

#### 7.2.1.2 Synthesis of metakaolin geopolymer
Geopolymer mixtures are synthesized using metakaolin as the precursor (primary source material). The activating solution is prepared using sodium hydroxide and fumed silica powder. The mixture is prepared by using $Na_2O$, $Al_2O_3$, $SiO_2$, and $H_2O$ producing geopolymer of high mechanical strength [9] and thermodynamically stable [10]. By adding fumed silica, NaOH, and $H_2O$ in a sealed steel container, an exothermic reaction occurs. Therefore, the reaction mixture is cooled before mixing with the metakaolin for 30 h to prepare an activating solution.

The solid-state $^{27}Al$ nuclear magnetic resonance (NMR) test is used to observe the change of 5- and 6-coordinated aluminum to 4-coordinated aluminum in metakaolin in geopolymer gel and also to know the relationship between conversion and setting. Liquid-state $^{29}Si$ NMR test is used to monitor the nanostructural evolution of aluminosilicates during geopolymerization and also know the relationship between evolution and setting behavior.

#### 7.2.1.3 Dissolution of aluminum (Al) during geopolymerization
The area of each deconvoluted peak is plotted versus time in $^{27}Al$ NMR tests. The amount of 4-coordinated aluminum increases, but that of the 5- and 6-coordinated amount of aluminum decreases with time. The changes are observed during the geopolymer reaction for the amount of aluminum, which indicates the dissolution of metakaolin [11]. This conversion is increasing quickly in the right side after mixing up to 15 h, then decreases and became immeasurable after 20 h. $^{27}Al$ is a quadrupolar nucleus; therefore, the quantitative interpretation was rare. $^{27}Al$ was used as an indicator for the conversion of aluminum from 5- and 6-coordinated aluminum to 4-coordinated aluminum. A higher portion of 4-coordinated aluminum with respect

to its own amount was detected [12]. They observed that dissolution of metakaolin to release aluminum is negligible after 20 h.

### 7.2.2 Synthesis of an inorganic polymer, that is, geopolymer

#### 7.2.2.1 Materials required
Pyrophyllite (mineral), aluminosilicate, and water ($H_2O$) are used.

#### 7.2.2.2 Synthesis of geopolymer
The main crystalline impurity present in pyrophyllite was quartz, whereas kaolinite and illite are present in small quantity. NaOH and water are mixed with aluminosilicate and then add sodium silicate solution to the mixture. The mixture is heated at 75 °C for 30 h and then dried at 80 °C.

#### 7.2.2.3 Characterization
The $^{27}$Al magic angle spinning (MAS) NMR spectrum of illite proves that it contains octahedral aluminum ($\delta$ = 6 ppm) and tetrahedral aluminum ($\delta$ = 80 ppm). The $^{29}$Si MAS NMR spectrum shows tetrahedral resonances at −100 ppm [13].

The $^{29}$Si MAS NMR spectrum contains a peak at $\delta$ 80 ppm [14]. When the pyrophyllite is dehydroxylated at 800 °C, the $^{27}$Al spectrum is unchanged, while the $^{29}$Si spectrum shows a peak at $\delta$ 88 ppm, which proves that the geopolymerized samples are derived from the unheated material. By comparing the dehydroxylated phases of client and pyrophyllite structure, they suggest that the crystalline phase of pyrophyllite undergoes geopolymerization [15].

### 7.2.3 Use of geopolymer cement for making the materials of green construction

#### 7.2.3.1 Materials required
Kaolinite, sodium silicate ($Na_2SiO_3$), silicon oxide ($SiO_2$), sodium oxide ($Na_2O$), aluminum oxide ($Al_2O_3$), water, and sodium hydroxide (NaOH) are used.

#### 7.2.3.2 Synthesis of geopolymer cement
Geopolymer cement is prepared by using $Na_2SiO_3$, kaolinite, and NaOH. When $Na_2SiO_3$ and NaOH are dissolved in $H_2O$, an exothermic reaction occurs. Then add metakaolinite to the sodium hydroxide and sodium silicate solutions, a reaction mixture is generated. The reaction mixture is stirred continuously at 56 °C for 8 h. Then filtered and dried the product.

### 7.2.3.3 Characterization
Sharp peaks were observed in X-ray diffraction (XRD) pattern between 22° and 48°, which indicates the presence of the amorphous phase. These sharp peaks were observed due to calcination and metakaolinite-alkali geopolymerization [16]. The formation of the amorphous phase (gel) indicates that the geopolymerization of kaolinitic soil occurs.

Scanning electron microscopic (SEM) analysis proves that the geopolymer gel is used as a binder material to the end products of geopolymer cement. The high amount of binder material indicates the high mechanical performance and stability of the produced geopolymer cement [17]. The layers of metakaolin layers reported by the SEM are distorted due to the calcination of kaolinite that occurred at 840 °C. SEM analysis shows the presence of microstructure of metakaolinite-based geopolymers. SEM image proves that geopolymeric gel is used to fill the gap between partially reacted metakaolinite layers.

## 7.2.4 Factors affecting the rate of geopolymeric gels

### 7.2.4.1 Material required
Metakaolin, alkaline silicate solution, and water are used.

### 7.2.4.2 Synthesis of geopolymer
A geopolymer sample is synthesized by mixing metakaolin and an alkaline silicate solution mechanically. After 20 min the sample is centrifuged to remove air entered into it. Samples of geopolymer that contain sodium cation are known as a Na-geopolymer. Samples of geopolymer that contain sodium and potassium are known as Na–K-geopolymer. Samples of geopolymer that contain potassium cations are known as a K-geopolymer.

### 7.2.4.3 Effect of water in geopolymer
Figure 7.1 shows $^2H$ MAS NMR spectra for Na-geopolymers of differing Si/Al ratios. Broadband was observed in the center due to the presence of aluminosilicate glasses and minerals [18]. As the concentration of Si/Al ratio increases, hydrogen mobility decreases. As the content of silica present in geopolymers increases, the size of water and hydroxide decreases. Broadband was observed due to the presence of water in the crystalline structures (Figure 7.2).

### 7.2.4.4 Role of alkali in aluminum incorporation
The peak is observed at −6.0 δ for all the samples of geopolymer and is due to the balance of the charge of sodium and aluminum.

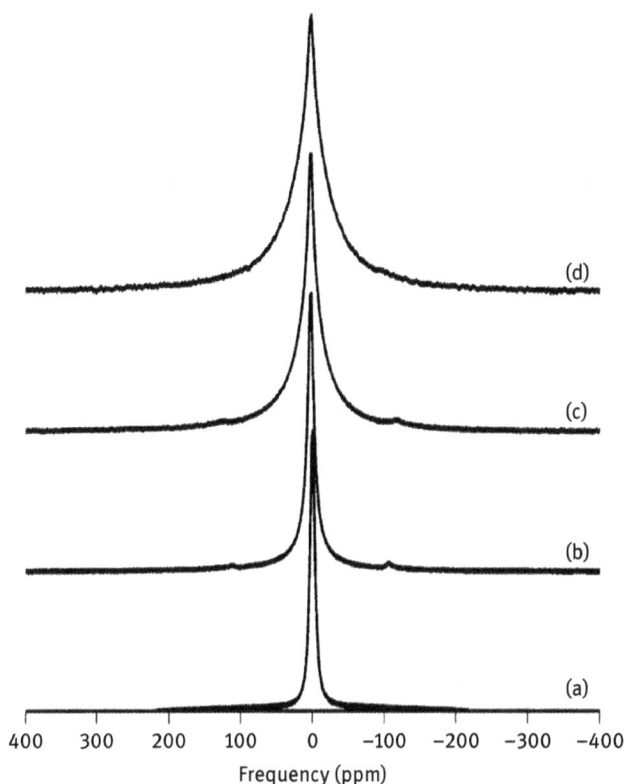

**Figure 7.1:** $^2$H MAS NMR spectra of Na-geopolymers with different ratios of Si/Al (a) 1.20, (b) 1.50, (c) 1.75, and (d) 2.20. Reproduced, with permission, P. Duxson, G. C. Lukey, F. Separovic, and J. S. J. van Deventer, Ind Eng Chem Res 2005, 44, 832–839.

## 7.2.5 Synthesis of geopolymers for construction and water purification

### 7.2.5.1 Materials required
Metakaolin, zeolitic, and NaOH were used.

### 7.2.5.2 Synthesis of metakaolin-zeolitic tuff geopolymer
The metakaolin-zeolitic tuff geopolymer is prepared using metakaolin, natural zeolitic tuff (JZ), $Na_2SiO_3$, and NaOH. When the activating solution is prepared by mixing $Na_2O$, $Al_2O_3$, $SiO_2$, NaOH, and $H_2O$, an exothermic reaction occurs. Therefore, the reaction mixture is cooled to prepare an activating solution. $Na_2SiO_3$ and NaOH are dissolved in $H_2O$ and mixed for 10 min mechanically. To produce required geopolymer, zeolitic tuff, and metakaolin, $Na_2SiO_3$ and NaOH were mixed.

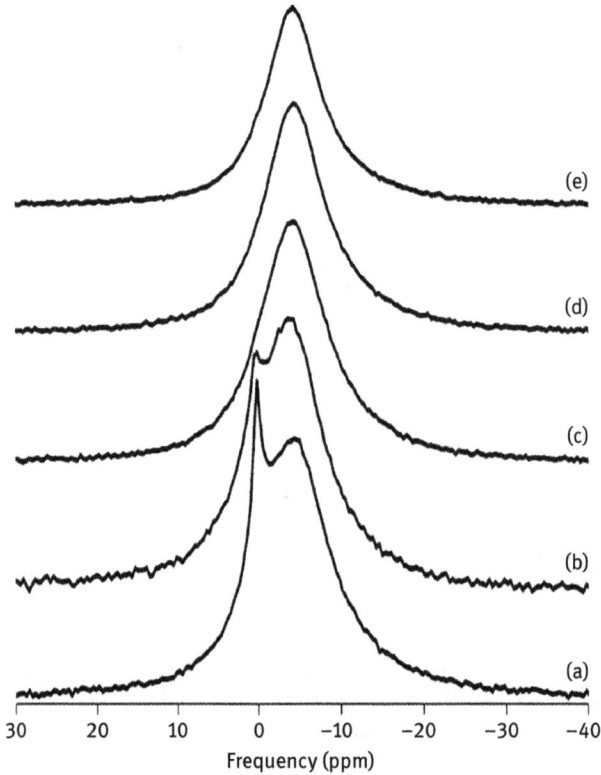

**Figure 7.2:** $^{23}$Na MAS NMR spectra of Na-geopolymers with different ratios of Si/Al (a) 1.20, (b) 1.50, (c) 1.75 (d) 1.95 (e) 2.20. Reproduced, with permission, P. Duxson, G. C. Lukey, F. Separovic, and J. S. J. van Deventer, Ind Eng Chem Res 2005, 44, 832–839.

### 7.2.5.3 Characterization

The broad, sharp peak is observed in XRD between 28° and 45° [19]. The SEM image prove that sodium aluminosilicate is used to fill the gap between partially reacted metakaolin sheets [20]. A bigger particle of the aluminosilicate particles indicates the larger pore area. A gel consisting of smaller particles indicates the smaller pore area [21].

### 7.2.6 Factors affecting the rate of geopolymers

### 7.2.6.1 Materials required

Fly ash (FA), NaOH, Na$_2$SiO$_3$, and sodium silicate are used.

### 7.2.6.2 Synthesis of geopolymer
Geopolymer is produced when NaOH was mixed with $Na_2SiO_3$ and fly ash.

### 7.2.6.3 Properties of geopolymer
The compressive strength of different NaOH concentration is reported. As the concentration of NaOH increases, the compressive strength of geopolymer increases [22–28].

### 7.2.6.4 Characterization
SEM micrographs prove that the size of the fly ash particle decreases with an increase in the concentration of NaOH [29, 30].

## 7.2.7 Geopolymer matrix

### 7.2.7.1 Materials required
$Na_2SiO_3$, NaOH, potassium hydroxide (KOH), $NaNO_3$, hydrogen chloride (HCl), $CH_3$-$COONH_4$, sodium aluminate, fly ash, Kembla blast furnace slag, kaolinite, K-Feldspar, and potassium silica are used. Polyethylene containers are used in this synthesis of geopolymer to prevent silicon contamination.

### 7.2.7.2 Synthesis of geopolymer sample
Fly ash is mixed either with kaolinite or with K-feldspar in a Kembla blast furnace slag, to get a geopolymer mixture [31, 32].

### 7.2.7.3 Characterization of geopolymeric matrixes
Matrixes containing kaolinite produce the highest mechanical strength.

In XRD, fly ash shows peaks between 25° and 45° due to the amorphous material, that is, silica [33]. Fly ash shows peaks between 50° and 60° because of the crystalline material, that is, mullet, hematite, and quartz [15, 34].

## 7.3 Applications

Geopolymerisation reactions open new applications, which already find applications in every industry, whether used pure, filled or reinforced. These applications include the automobile and aviation industries, non-ferrous smelter and metallurgy, civil engineering, the concrete and cement sectors, the ceramics and plastics industries, the waste management industry. Geopolymer composites have three major characteristics that render them inferior to glass matrix composites, polymers and organic composites. Rock-based geopolymer cement is suitable for economic systems, such as continuous encapsulation of radioactive material as well as other dangerous waste, poisonous materials as well as seals, cappings, obstacles and other buildings required for remedying the poisonous waste confinement sites.[35-48]

## 7.4 Conclusion

Metakaolin-zeolitic tuff geopolymer is used for the removal of micropollutants and toxic products. Geopolymer is the best material for the protection of environment because it reduces the production of waste.

## References

[1]    Davidovits, J. Review: Geopolymers: Ceramic-like inorganic polymers, J. Ceram. Sci. Technol., 2017, 08(3), 335–350.

[2]    Kim, D., Lai, H.T., Chilingar, G. V., & Yen, T.F. Geopolymer formation and its unique properties, Environ. Geol., 2006, 51, 103–111.

[3]    Duxson, P., Fernandez-Jimenez, A., Provis, J. L., Lukey, G. C., Palomo, A., & van Deventer, J.S.J. Geopolymer technology: the current state of the art, J. Mater. Sci., 2007, 42, 2917–2933.

[4]    Duxson, P., Provis, J.L., Lukey, G.C., & van Deventer, J.S.J. The role of inorganic polymer technology in the development of green concrete, Cement Concrete Res., 2007, 37, 1590–1597.

[5]    Alshaaer, M., El-Eswed, B., Yousef, R.I., Khalili, F., & Rahier, H. Development of functional geopolymers for water purification and construction purposes, J. Saudi Chem. Soc., 2016, 20, S85–S92.

[6]    El- Eswed, B., Yousef, R.I., Alshaaer, M., Khalili, F., & Khoury, H. Alkali solid-state conversion of kaolin and zeolite to effective adsorbents for removal of lead from aqueous solution, Desalin. Water Treat., 2009, 8, 124–130.

[7]    Li, L., Wang, S., & Zhu, Z. Geopolymeric adsorbents from fly ash for dye removal from aqueous solution, J. Colloid. Interface. Sci., 2006, 300, 52–59.

[8]    Wang, S., Li, L., & Zhu, Z.H. Solid-State conversion of fly ash to effective adsorbents for Cu removal from wastewater, J. Hazard. Mater., 2007, 139, 254–259.

[9]    Duxson, P., Mallicoat, S.W., Lukey, G.C., Kriven, W.M., & van Deventer, J.S.J. The effect of alkali and Si/Al ratio on the development of mechanical properties of metakaolin-based geopolymers, Colloids Surf. A, 2007, 292, 8–20.

[10]   Criado, M., Fern andez-Jim Enez, A., Palomo, A., Sobrados, I., & Sanz, J. Effect of the $SiO_2/Na_2O$ ratio on the alkali activation of fly ash. Part II: 29Si MAS-NMR survey, Microporous. Mesoporous. Mater., 2008, 109, 525–534.

[11]   Provis, J.L., & Van Deventer, J.S.J. Geopolymers: Structures, processing, properties and industrial applications. Boca Raton, FL:CRC Press, Inc, 2009, 72–88.

[12]   Fyfe, C.A., Bretherton, J.L., & Lam, L.Y. Detection of the invisible aluminum and characterization of the multiple aluminum environments in zeolite USY by high-field solid-state NMR, Chem. Commun., 2000, 0(17), 1575–1576.

[13] Barbosa, V.F.F., MacKenzie, K.J.D., & Thaumaturgo, C. Synthesis, and characterization of materials based on inorganic polymers of alumina and silica: Sodium polysialate polymers, Int. J. Inorg. Mater., 2000, 2, 309–317.

[14] Brew, D.R.M., & MacKenzie, K.J.D. Geopolymer synthesis using silica fume and sodium aluminate, J. Mater. Sci., 2007, 42, 3990–3993.

[15] El-Eswed, B.I., Yousef, R.I., Alshaaer, M., Hamadneh, I., Al-Gharabli, S.I., & Khalili, F. Stabilization/solidification of heavy metals in kaolin/zeolite based geopolymers, Inter. J. Miner. Process., 2015, 137, 34–42.

[16] Khale, D., & Chaudhary, R. Mechanism of geopolymerization and factors influencing its development: A review, J. Mater. Sci., 2007, 42, 729–746.

[17] Schaller, T., & Sebald, A. One- and two-dimensional 1H magic-angle spinning experiments on hydrous silicate glasses. Solid. State, Nucl. Mag., 1995, 5, 89–102.

[18] Alshaaer, M., Cuypers, H., Rahier, H., & Wastiels, J. Evaluation of a low-temperature hardening inorganic phosphate cement for construction and industrial applications, Cem. Conc. Res., 2011, 41, 38–45.

[19] Rahier, H., Simons, W., Biesemans, M., & Van Mele, B. Low-temperature synthesized aluminosilicate glasses: Part III influence of the composition of the silicate solution on production, structure, and properties, J. Mater. Sci., 1997, 32, 2237–2247.

[20] Setzer, C., van Essche, G., & Pryor, N. Title of the article, Schuth, F., Sing, K.S.W. Weitkamp, J., Eds., Handbook of Porous Solids, Wiley-VCH, Weinheim, 2002, Vol. 3, 1543.

[21] Posi, P., Teerachanwit, C., Tanutong, C., Limkamoltip, S., Lertnimoolchai, S., Sata, V., & Chindaprasirt, P. Lightweight geopolymer concrete containing aggregate from recycle lightweight block, Mater. Des., 2013, 52, 580–586.

[22] Onutai, S., Wasanapiarnpong, T., Jiemsirilers, S., Wada, S., & Thavorniti, P. Effect of sodium hydroxide solution on the properties of geopolymer based on fly ash and aluminium waste blend, Suranaree. J. Sci. Technol., 2012, 21, 9–14.

[23] Mustafa Al Bakri, A.M., Kamarudin, H., Bnhussain, M., Khairul Nizar, I., Rafiza, A.R., & Zarina, Y. Microstructure of different NaOH molarity of fly ash based green polymeric cement, J. Eng. Technol. Res., 2011, 3, 44–49.

[24] Palomo, A., Varela, M.T.B., Granizo, M.L., Puertas, F., Vazquez, T., & Grutzeck, M.W. Chemical stability of cementitious materials based on metakaolin, Cem. Concr. Res., 1999, 29, 997–1004.

[25] Thokchom, S., Ghosh, P., & Ghosh, S. Effect of water absorption, porosity, and sorptivity on the durability of geopolymer mortars, ARPN J. Eng. Appl. Sci., 2009, 4, 28–32.

[26] Pimraksa, K., Chindaprasirt, P., Rungchet, A., Sagoe-Crentsil, K., & Sato, T. Lightweight geopolymer made of highly porous siliceous materials with various $Na_2O/Al_2O_3$ and $SiO_2/Al_2O_3$ ratios, Mater. Sci. Eng. A, 2011, 528, 6616–6623.

[27] Bakri, A.M.M.A., Kamarudin, H., Bnhussain, M., Nizar, I.K., & Mastura, W.I.W. Mechanism and chemical reaction of fly ash geopolymer cement- a review, Asian. J. Sci. Res., 2011, 5, 247–253.

[28] Hanjitsuwan, S., Hunpratub, S., Thongbai, P., Maensiri, S., Sata, V., & Chindaprasirt, P. Effects of NaOH concentrations on physical and electrical properties of high calcium fly ash geopolymer paste, Cem. Concr. Comp., 2014, 45, 9–14.

[29] Van Jaarsveld, J.G.S., Van Deventer, J.S.J., & Lorenzen, L. The potential use of geopolymeric materials to immobilise toxic metals: Part II material and leaching characteristics, Miner. Eng., 1999, 12, 75–91.

[30] Phair, J.W., & Van Deventer, J.S.J. Effect of the silicate activator on the leaching and material characteristics of waste-based geopolymers, Miner. Eng., 2001, 14, 289–304.

[31] Phair, J.W., Van Deventer, J.S.J., & Smith, J.D. Mechanism of polysialation in the incorporation of zirconia into fly ash based geopolymers, Ind. Eng. Chem. Res., 2000, 39, 2925–2934.

[32] Clark, B.A., & Brown, P.W. The formation of calcium sulfoaluminate hydrate compounds Part II, Cem. Concr. Res., 2000, 30, 233–240.

[33] Skibsted, J., Henderson, E., & Jakobsen, H.J. Characterisation of calcium aluminate phases in cements by $^{27}$Al MAS NMR spectroscopy, Inorg. Chem., 1993, 32, 1013–1027.

[34] Alshaaer, M. Two-phase geopolymerization of kaolinite-based geopolymers, Appl. Clay. Sci., 2013, 86, 162–168.

[35] Alshaaer, M., El-Eswed, B., Yousef, R.I., Khalili, F., & Khoury, H. Low-cost solid geopolymeric material for water purification. Environmental issues and waste management technologies in the materials and nuclear industries XII: Ceramic Transactions, published by Wiley, 2009, 207.

[36] Yousef, R., El-Eswed, B., Alshaaer, M., Khalili, F., & Khoury, H. The influence of using Jordanian natural zeolite on the adsorption, physical, and mechanical properties of geopolymers products, J. Hazar. Mater., 2009, 165, 379–387.

[37] Abdullah, M.M.A., Kamarudin, H., Mohammed, H., Khairul Nizar, I., Rafiza, A.R., & Zarina, Y. The relationship of NaOH molarity, Na$_2$SiO$_3$/NaOH ratio, fly ash/alkaline activator ratio, and curing temperature to the strength of fly ash-based geopolymer, Adv. Mater. Res., 2011, 328, 1475–1482.

[38] Subramaniam, V., Ngan, M.A., May, C.Y., & Sulaiman, N.M.N. Environmental performance of the milling process of Malaysian palm oil using the life cycle assessment approach, Am. J. Environ. Sci., 2008, 4(4), 310–315.

[39] Duxson, P., Fernandez-Jimenez, A., Provis, J.L., Lukey, G.C., Palomo, A., & Van Deventer, J.S.J. Geopolymers: An environmental alternative to carbon dioxide producing ordinary Portland cement, J. Mater. Sci., 2007, 42, 2927–2933.

[40] Peric, J., Trgo, M., & Medvidovic, N.V. Removal of zinc, copper and lead by natural zeolite-a comparison of adsorption isotherms, Water. Res., 2004, 38, 1893–1899.

[41] Gunasekaran, S., & Anbalagan, G. Thermal decomposition of natural dolomite, Bull. Mater. Sci., 2007, 30(4), 339–344.

[42] Zain, H., Abdullah, M.M.A.B., Hussin, K., Ariffin, N., & Bayuaji, R. Review on various types of geopolymer materials with the environmental impact assessment. MATEC Web of Conferences 2017, 97.

[43] Kranjc, A. Baltazar hacquet (1739/40–1815), the pioneer of karst geomorphologists, Acta Carsolog., 2006, 35(2), 163–168.

[44] Kosmatka, S.H., Kerkhoff, B., & Panarese, W.C. The design and control of concrete mixtures. Portland Cement Association: Skokie, Illinois, 2002, 1–358.

[45] Kroehong, W., Sinsiri, T., & Jaturapitakkul, C. Effect of palm oil fuel ash fineness on packing effect and Pozzolanic reaction of blended cement paste, Procedia. Eng., 2011, 14, 361–369.

[46] Bergamonti, L., Alfieri, I., Lorenzi, A., Predieri, G., Barone, G., Gemelli, G., Mazzoleni, P., Raneri, S., Bersani, D., & Lottic, P.P. Nanocrystalline TiO$_2$ coatings by sol–gel: Photocatalytic activity on Pietra di Noto biocalcarenite, J. Sol-Gel Sci. Technol., 2015, 75, 141–151.

[47] Zarina, Y., Abdullah, M.M.A., Kamarudin, H., Khairul Nizar, I., & Rafiza, A.R. Review on the various ash from palm oil waste as geopolymer material, Rev. Adv. Mater. Sci., 2013, 34, 37–43.

[48] Cocca, M., D'Arienzo, L., D'Orazio, L., Gentile, G., & Martuscelli, E. Polyacrylates for conservation: chemico-physical properties and durability of different commercial products, Polym. Test., 2004, 23, 333–342.

# 8 Advanced characterization methods of inorganic polymers

**Abstract:** The characterization of polymers is aimed at improving the performance of polymeric materials. Many characterization techniques increase the properties of materials such as thermal stability, strength, impermeability, and optical properties. The techniques of characterization are mainly used for the determination of the mass of molecules, the structure of molecules, its morphology, and thermal and mechanical properties.

**Keywords:** characterization, spectroscopy, nuclear magnetic resonance (NMR), thermal microscopy, XRD

## 8.1 Introduction

The characterization of polymer comes under the analytical branch of polymer science. This branch is concerned for the polymer characterization at various levels, including the composition and structure (including defects) of materials. This characterization is precious to synthesize new polymers and also for improving the performance of the product. The characterization of the polymer involves the distribution of molecular weight, structure of the molecule, polymer morphology, and thermal, mechanical, and optical properties. Many properties of materials are linked to these characterization techniques. These techniques also include refinement and the development of analytical methods with statistical models, helping us to understand phase transitions and phase separation of polymers, which is a complex and multifaceted process.

## 8.2 Infrared and Raman spectroscopy

Infrared (IR) spectroscopy can be used to characterize long-chain polymers because the IR active groups present along the chain absorb as if each was a localized group in a simple molecule. IR arises due to the interaction between the frequency of the IR incident radiation and that of a particular vibrational mode, whereas in Raman effect inelastic scattering of light occurs when a monochromatic light irradiates on the molecule. Raman scattering analyzes only thick polymer samples, whereas IR analyzes the samples of very thin films.

https://doi.org/10.1515/9781501514609-009

## 8.3 Fluorescence spectroscopy

The fluorescence spectroscopy requires a fluorescent probe at lowest concentration [1, 2]. Zammarano et al. reported the formation of polymer nanocomposites [3, 4].

Yu et al. studied about the synthesis of the extracellular polymeric substances (EPS), which was extracted from aerobic and anaerobic sludges in the treatment of wastewater [5]. Sorgjerd et al. studied the synthesis of pre-fibrillar oligomer aggregates of transthyretin [6]. Yu et al. reported the formation of a continuous layer of biofilm having a thickness of 20–100 μm [7, 8].

Ji et al. investigated the preparation of supramolecular cross-linked network by using conjugated polymer and a bis-ammonium salt cross-linker. Dibenzo [24] crown-8 acts as a multiple fluorescent sensor in this preparation [9]. Wu et al. [10] reported the synthesis of conjugated polymer dots [10].

## 8.4 NMR uses solid-state methods

High-resolution NMR is one of the important tools to determine the microstructure of polymers in solution. Using this technique, the extensive molecular motion is reduced; therefore, long-range interactions are not possible, which allows only the short-range effects. Not only is the local field acting in the nucleus altered by the environment, but it is also sensitive to molecular motion and it has been observed that as the molecular motion within a sample increases, but the resonance lines also become narrower. Chain-end group signal provides an estimation of the average molecular weight. Because of polymer phase transition experiments on various temperature ranges, they exhibit some changes. Various 2D solid-state NMR techniques are present today and are regularly used for material in the study [11–24].

Dorkoosha et al. reported the synthesis of superporous hydrogel (SPH) and furthermore characterized by employing 2D NMR [25–28]. Bertmer et al. reported the synthesis of biodegradable polymer networks through UV curing, and characterization was done by solid-state NMR [29]. Gussoni et al. investigated the use of proton NMR imaging and relaxometry for determining the structure of elastomer polymers [30]. Rio et al. studied the structure of the lignin in the straw of wheat [31]. Agarwal et al. reported the use of colloidal particles of poly(4-vinyl pyridine) and silica for finding the interactions of molecules between the polymer and silica phases [32]. Figure 8.2 depicts the $^{31}$P-NMR spectrum of aniline-substituted polyphosphazene having signals at 2.3 and –15 ppm, which are attributed to the phosphorus atoms in different environments with aniline and chlorine, respectively [33] [Figure 8.1].

Zhang and coworkers have investigated an important reaction conditions like temperatures (Figure 8.2), time (Figure 8.3), and concentrations (Figure 8.4) on poly (bis(phenoxy)phosphazene) using various $^{31}$P-NMR spectra to correlate specific

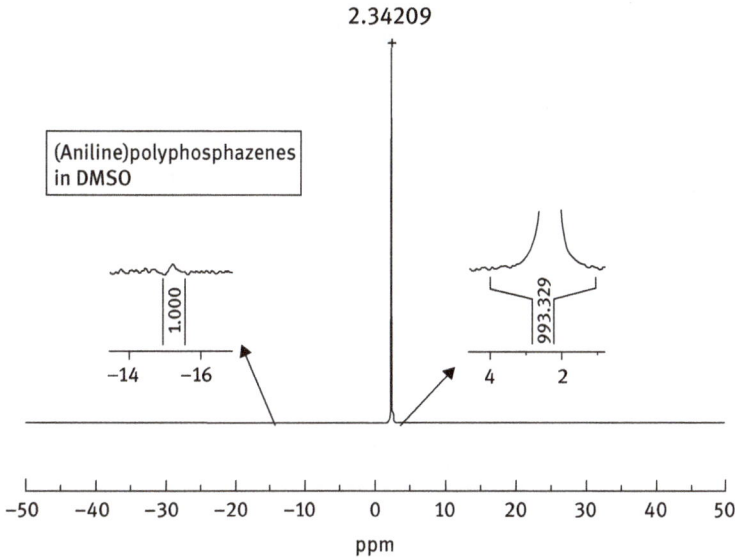

**Figure 8.1:** $^{31}$P-NMR spectra of aniline polyphosphazene [33].

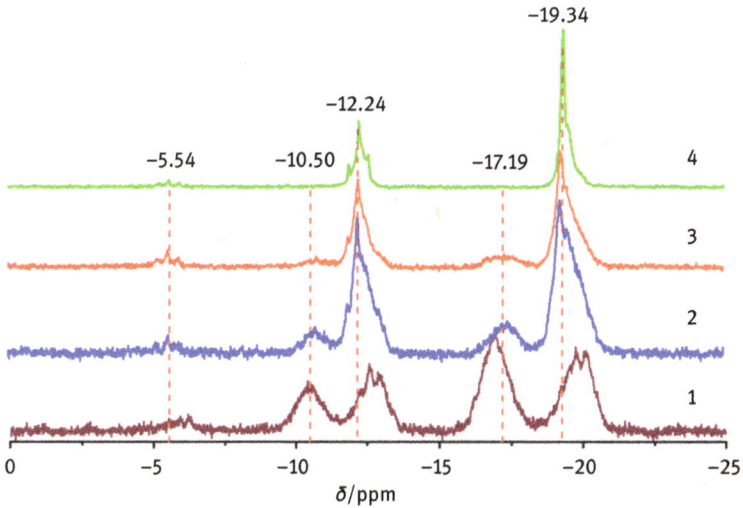

**Figure 8.2:** $^{31}$P-NMR spectra of the substitution reaction products, where the reaction of poly (dichlorophosphazene) (PDCP) with sodium phenoxide was carried out in THF at different temperatures: (1) 30 °C, (2) 60 °C, (3) 80 °C, and (4) 100 °C [34].

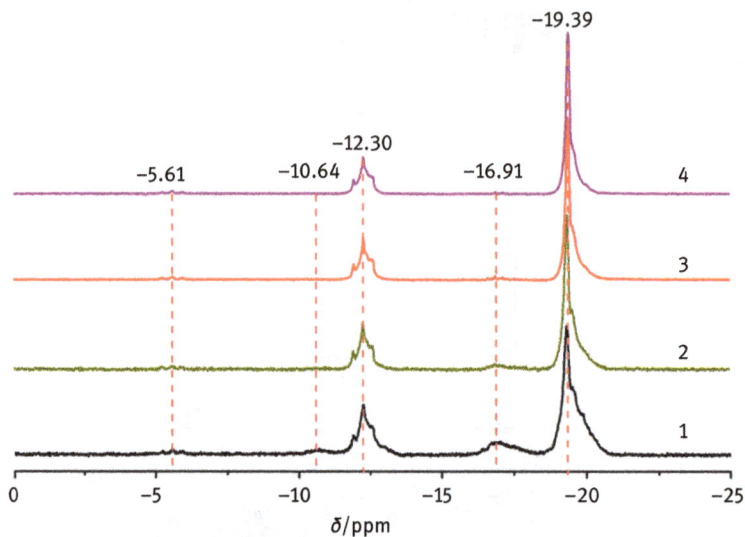

**Figure 8.3:** $^{31}$P-NMR spectra of substitution reaction products, where PDCP reacted with sodium phenoxide in THF for (1) 6 h, (2) 12 h, (3) 18 h, and (4) 24 h, respectively [34].

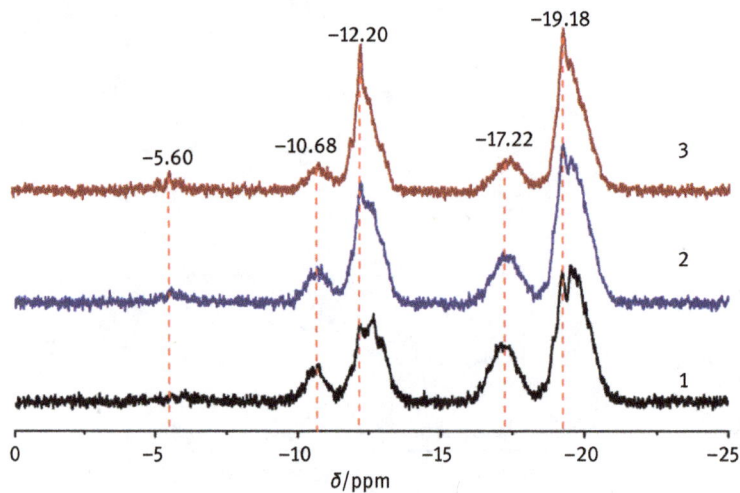

**Figure 8.4:** $^{31}$P-NMR spectra of the substitution reaction products, where PDCP reacted with sodium phenoxide at different PDCP concentrations in THF: from (1) 1 g/100 mL, to (2) 2 g/100 mL, and to (3) 3 g/100 mL [34].

structural defects. It is concluded from Figure 8.2 that with the increase in tempera-ture, the peak width decreases with increasing temperature [34].

## 8.5 Photoacoustic spectroscopy

Photoacoustic spectroscopy (PAS) uses waves produced by materials, that is, acoustic waves, exposing them to light for concentration measurement. PAS combines heat measurements with optical microscopy owing to its unique nature. PAS is generally used for gases, but research is continuing to use PAS efficiently for solid and liquid samples. PAS measures directly by using an internal heat instead of the effects of the light on the surroundings, making the measurement accurate and useful for sensitive detectors. The interest in PAS came into limelight in 1880 after Alexander Graham Bell discovered the acoustic effect [35].

## 8.6 Microscopy

Scanning electron microscopy (SEM) is the main advantage of electron microscopy, which has much-improved resolution compared to that of light microscopy. In a SEM, a finely focused electron beam is scanned across the specimen, and scattered elec-trons emitted at each point are collected by appropriate detectors forming an image. Backscattered electrons are high-energy electrons that have been elastically scattered by the interaction of the incident beam with the nucleus. The energy of these elec-trons is comparable to that of the incident beam. SEM instruments have resolutions better than 5 nm and are useful for the characterization of surfaces and determination of surface topography. Contrary to transmission electron microscopy (TEM), little sample preparation is required. In addition to detecting backscattered and secondary electrons, SEM instruments offer information on the sample elemental composition when using X-ray detectors. It is already reported that the ejection of electrons from an atom is accompanied by an emission of X-rays. The X-ray spectrum that is pro-duced is a characteristic feature of any given element, and by measuring the energy or the wavelength of the X-rays that are produced, its identification is possible. AFM (atomic force microscopy) and STM (scanning tunneling microscopy) offer a means of obtaining three-dimensional images of polymer surfaces. These technologies offer ad-vantages compared to electron microscopy because $N$ sample preparation is required such as coating or microtoming. These are a scanned-proximity probe microscopy technique that measures a local property, using a probe or tip placed close to the sample. STM probe is placed a few angstroms above the surface to measure a tunnel-ing current between the sample and the tip of the probe, whereas the AFM probe is in contact with the sample surface. STM is limited to studies of conducting or metal coating polymers, whereas the AFM is more useful in polymer surface studies. AFM measures attractive or repulsive forces between a probe and a sample. AFM is widely used in the analysis of polymer surfaces, such as morphology and molecular structure of crystalline and oriented polymers, block copolymers, and polymer blends. TEM is used to analyze very thin samples less than 100 nm thick, provided that the specimen has structural features that scatter electrons in different amounts.

AFM is a very important tool in material science, which gives information about surface irregularities and structures with superior spatial resolution. In AFM, a repulsive force between the tip and the sample provides more accurate surface analysis.

An analog of AFM, STM is another characterization tool for the high-resolution image of the surfaces of an insulator. AFM is a multifunctional technique, which can be used for the characterization of topography, adhesion, and mechanical properties.

Contrast and resolution are two important parameters of microscopy to examine measurement and morphology. Contrast mechanism of AFM is based on the mechanical interaction between the tip and sample, that's why it is used to provide morphology on polymers. It is quite different than the contrast mechanism between electrons and the sample in SEM or TEM. Contrast required for SEM and TEM can be difficult to obtain for polymers since there might not be much chemical differentiation between the materials. Most common modes of various AFM modes are based on mechanical contrast, which is called as phase imaging. Various mechanical properties such as modulus, adhesion, and viscoelasticity can be measured with impressive nanometer lateral resolution and subnanometer vertical resolution qualitatively with the help of AFM. Phase imaging is a dynamic technique where the vibration of the oscillator at resonance measures the phase lag. The phase lag measures the stage slack between the drive and reaction. This phase lag is attributed due to the convolution of material properties, including adhesion, dissipation, stiffness, and viscoelasticity for better qualitative information with good contrast. Hybrid materials based on both inorganic and organic phases are fabricated by sol–gel inorganic polymerization for various industrial and military applications. Herein, AFM imaging of hybrid materials provides qualitative information of organic and inorganic polymers and also describe boundary region between the inorganic and organic phase in hybrid material.

## 8.7 Thermal analysis

Under the controlled temperature program, various physical properties are functions of temperature at the same time by the help of various techniques. A wide range of thermal analysis technique has been developed such as thermogravimetry (TG), differential scanning calorimetry (DSC), differential thermal analysis (DTA), and dynamic mechanical analysis (DMA). TG monitors the change in the mass of material during a controlled temperature range. It can be used with FT-IR (Fourier-transform infrared) or pyrolysis of GC-MS to identify and quantify additives such as organic and inorganic fillers, plasticizers, antioxidants, stabilizers, UV absorbers, nucleating agents, and lubricants [36]. The derivative curve of TG, also called as DTG, can be used to improve the determinations of onset and endset points of decomposition of the low or multilevel polymer matrix. DTA monitors the temperature difference between specimen and reference material, whereas DSC heats reference and specimen separately and measures the heat flow between the materials. Thermal characterization was a technique in which the property of the sample was monitored against time or temperature. FT-IR technique was used to study the structure of resins and thermal stability was measured using TG [37, 38]. Barontini et al. have studied the decomposition of products containing BFRs brominated flame retardants [39].

DSC and DTA techniques are used to determine various physical processes like melting, crystallization, and crystalline disorientation [40]. DSC helps us to understand the crystallinity, cross-linking, activation energy, kinetic parameters, entropy change, and heat of polymerization [41].

A maximum polymer study is done with dynamic TG. The sample loses its weight slowly with the rising temperature. At a certain temperature, the weight loss becomes zero [42–44]. Xie et al. [45] reported that the thermal stability of the organically exchanged montmorillonite is affected by the chemical variation like chain length of alkyl group and unsaturation of organic modifiers. Carrasco et al. [46] studied the chemical and mechanical properties of poly(lactic acid). Cerqueira and coworkers [47] studied the structure of galactomannans from *Gleditsia triacanthos*, *Caesalpinia pulcherrima*, and *Adenanthera pavonina* by using some nonconventional sources. Verploegen et al. [48] investigated that the temperature affects the morphology of poly(3-hexylthiophene)–phenyl-C61-butyric acid methyl ester thin film bulk heterojunction blends.

## 8.8 X-ray diffraction and single-crystal XRD

X-ray diffraction (XRD) is a very useful method to evaluate the structure and composition of organic and inorganic crystals. The polymers and fibers and their blends may consist of both crystalline and amorphous phases. It is a nondestructive and rapid analytical tool to examine the amount of crystalline content present in polymers and fibers. The crystalline part easily diffracted by high-energy X-ray light ultimately provides diffractogram. The degree of crystallinity, Miller indices, and other crystallographic data of the materials can be determined by peak integration by Bragg's law.

The polymer is a type of soft matter that can easily deform and not easy to crystallize. However, XRD is used to recognize different types of polymer phases (crystalline, semicrystalline, and amorphous), identification of polymers, and measuring crystallinity.

The most well-known test technique for acquiring a detailed structure of a particle, which permits resolutions of individual atoms is single-crystal XRD, and this is performed by analyzing the array of many identical molecules. Numerous perfect materials like organometallic molecules, proteins, and polymers under specific condition can easily solidify into its crystalline form and adopt three-dimensional structures. When the X-ray beam is incident on a single crystal, then these specific radiations interact with an electron of an atom, which ultimately produces a particular pattern. As the crystal is rotated in the X-ray beam, multiple images are recorded with the help of the detector.

Perfect single crystals are in fact composed of translationally ordered molecules or atoms, and they determine the neighborhood lattice. Although individual dislocations disrupt the neighborhood symmetry, the worldwide translational symmetry is actually maintained. The development of an excellent polymer crystal, with the lattice realized by a one-time orientation [49] with no chain folding, takes conspicuously a bit more time than that needed for the improvement of little molecule crystals. The quantity of chain folding in the polymer crystals reflects metastability and is actually driven by the crystal growth rate [50]. Polymer crystals possess different chain ordering and conformational entropy, which results in different melting temperatures [51, 52]. Nevertheless, for the polymers, the entropy alteration differs from a single unit to a completely different one [53]. Despite metastability, all of the polymers within the single crystals have precisely the same orientation. The orientation of the corresponding device cell is still developed by the nucleation situation and most of the ensuing polymers are actually incorporated based on the very same unit cell parameters. Polymer chains can crystallize into different metastable states with various quantities of chain folding.

Ma et al. [54] obtained different kinds of morphologies of poly(ε-caprolactone) (PCL) crystals comprising finger-like spherulitic and dendritic by altering the solution concentration. The effect of interface and surface of the crystal framework along with the crystal orientation is examined from the viewpoints of the film thickness, crystallization temperature, and then interracial interaction [55]. Reiter et al. performed work that is actually outstanding on small film crystallization and ultrathin polymers.

The single crystals of poly(ethene) (PE) depicted lozenge or perhaps truncated lenticular practice and lozenge shapes in a number of solvents [56–58]. For PE, PCL, and other polymers, truncations have also been recorded as an increased truncation with crystallization temperature [59–64]. Furthermore, triangular and lozenge single crystal practices of the poly(L lactic acid) along with poly(D lactic acid) have been developed in unique crystallization conditions [65, 66]. Asymmetric curvature of crystal growth faces in PE oligomers was also studied by Ungar and coworkers [67].

Only one of the primary theories describing the improvement of PSCs from dilute answer was provided by Lauritzen and Hoffman [68].The Hoffman concept, based on the distant relative rates of nucleation and spreading, found three different types of development regimes: regimes I, II, and III [69]. It is demonstrated that the first lamellar thickness is handled by entropic barriers rather than enthalpic elements considered in the Lauritzen–Hoffman's theory [70, 71].

The chain diffusion coefficients in the answer crystallized species had been pushed by Yao et al. [72]. Single molecule fluorescence microscopy and molecule tracking tactics had been set on to master the diffusion of single polyethylene oxide (PEO) chains in its monolayers on solid substrates, before and after the crystallization process. The results showed a great correlation between the crystallization method as well as chain diffusion [73]. The modeling of transient nucleation in the isothermal thickening of polymer lamellar crystals was substantiated by Liu and coworkers [74]. Linear lateral growth, lamellar thickening, and slipping of monomers along chain guidance had been exclusively discovered in the larger scale, and long-time molecular dynamic simulations were carried out by Luo and Sommer [75].

## 8.9 Conclusion

The spectroscopic characterization such as IR, Raman, fluorescence, and NMR and thermal characterization like TG, DSC, and DTA of polymeric materials have gained considerable attention for the development of new kinds of polymers.

## References

[1]   Bokobza, L. Investigation of local dynamics of polymer chains in the bulk by the excimer fluorescence technique, Prog. Polym. Sci., 1990, 15, 337–360.
[2]   George, G. A. Characterization of solid polymers by luminescence techniques, Pure Appl. Chem., 1985, 57, 945–954.
[3]   Zammarano, M., Maupin, P. H., Sung, L. P., Gilman, J. W., McCarthy, E. D., Kim, Y. S., & Fox, D. M. Revealing the interface in polymer nanocomposites, ACS Nano., 2011, 5, 3391–3399.
[4]   Maupin, P. H., Gilman, J. W., Harris, R. H., Bellayer, S., Bur, A. J., Roth, C., Murariu, M., Morgan, A. B., & Harris, J. D. Optical probes for monitoring intercalation and exfoliation in melt-processed polymer nanocomposites, Macromol. Rapid Commun., 2004, 25, 788–792.
[5]   Sheng, G. P., & Yu, H.Q. Characterization of extracellular polymeric substances of aerobic and anaerobic sludge using three-dimensional excitation and emission matrix fluorescence spectroscopy, Water Res., 2006, 40, 1233–1239.
[6]   Lindgren, M., Sörgjerd, K., & Hammarström, P. Detection, and characterization of aggregates, prefibrillar amyloidogenic oligomers, and protofibrils using fluorescence spectroscopy, Biophys. J., 2005, 88(6), 4200–4212.
[7]   Mantouvalou, I., Malzer, W., Schaumann, I., Lühl, L., Dargel, R., Vogt, C., & Kanngiesser, B. Reconstruction of thickness and composition of stratified materials by means of 3D micro X-ray fluorescence spectroscopy, Anal. Chem., 2008, 80, 819–826.
[8]   Yu, G. H., Tang, Z., Xu, Y. C., & Shen, Q. R. Multiple fluorescence labeling and two dimensional FTIR–$^{13}$C NMR hetero-spectral correlation spectroscopy to characterize extracellular polymeric substances in biofilms produced during composting, Environ. Sci. Technol., 2011, 45, 9224–9231.
[9]   Ji, X., Yao, Y., Li, J., Yan, X., & Huang, F. A. Supramolecular cross-linked conjugated polymer network for multiple fluorescent sensing, J. Am. Chem. Soc., 2013, 135, 74–77.

[10] Wu, C., Szymanski, C., Cain, Z., & McNeill, J. Conjugated polymer dots for multiphoton fluorescence imaging, J. Am. Chem. Soc., 2007, 129, 12904–12905.

[11] Battiste, J., & Newmark, R. A. Applications of F-19 multidimensional NMR, Prog. Nucl. Magn. Reson. Spectrosc., 2006, 48, 1–23.

[12] Newmark, R. A., Battiste, J. L., & Koivula, M. I. Abstract Paper Am. Chem. Soc. 221, 286-POLY (2001).

[13] Ramamoorthy, A., Wu, C. H., & Opella, S. J. Experimental aspects of multidimensional solid-state NMR correlation spectroscopy, J. Magn. Reson., 1999, 140, 131–140.

[14] Rocha, J., Morais, C.M., & Fernandez, C. In new techniques in solid-state NMR, 1st, ed. by, J. Klinowski, Springer, Berlin, 2005.

[15] Fayon, F., King, I.J., Harris, R.K., Evans, J.S.O., & Massiot, D.C.R. Application of the through bond correlation NMR experiment to the characterization of crystalline and disordered phosphates, Chim, 2004, 7, 351–361.

[16] O'Dell, L. A., Abou Neel, E. A., Knowles, J. C., & Smith, M. E. Identification of phases in partially crystallized Ti-, Sr- and Zn-containing sodium calcium phosphates by two-dimensional NMR, Mater. Chem. Phys., 2009, 114, 1008–1015.

[17] O'Dell, L. A., Guerry, P., Wong, A., Abou Neel, E. A., Pham, T. N., Knowles, J. C., Brown, S. P., & Smith, M. E. Quantification of crystalline phases and measurement of phosphate chain lengths in a mixed phase sample by P-31 refocused INADEQUATE MAS NMR, Chem. Phys. Lett., 2008, 455, 178–183.

[18] Chisholm, M. H., Iyer, S. S., McCollum, D. G., Pagel, M., & Werner-Zwanziger, U. Microstructure of poly(lactide). Phase-sensitive HETCOR spectra of poly(meso-lactide), poly(rac-lactide), and atactic poly(lactide), Macromolecules, 1999, 32, 963–973.

[19] Hou, S. S., Bonagamba, T. J., Beyer, F. L., Madison, P. H., & Schmidt-Rohr, K. Clay intercalation of poly(styrene-ethylene oxide) block copolymers studied by multinuclear solid-state NMR, Macromolecules, 2003, 36, 2769–2776.

[20] Luliucci, R., Taylor, C., & Hollis, W. K. $^{1}H/^{29}Si$ cross polymerization NMR experiments of silica reinforced polydimethylsiloxane elastomers: probing the polymer-filler interface, Magn. Reson. Chem., 2006, 44, 375–384.

[21] Sroka-Bartnicka, A., Olejniczak, S., Ciesielski, W., Nosal, A., Szymanowski, H., Gazicki-Lipman, M., & Potrzebowski, M. Solid-state NMR study and density functional theory (DFT) calculations of structure and dynamics of poly(p-xylenes), J. Phys. Chem., B, 2009, 113, 5464–5472.

[22] Brown, S. P., Lesage, A., Elena, B., & Emsley, L. Probing proton-proton proximities in the solid state: high-resolution two-dimensional $^{1}H$-$^{1}H$ double-quantum CRAMPS NMR spectroscopy, J. Am. Chem. Soc., 2004, 126, 13230–13231.

[23] Coelho, C., Azais, T., Bonhomme-Coury, L., Laurent, G., & Bonhomme, C. Efficiency of the refocused $^{31}P$–$^{29}Si$ MAS-J-INEPT NMR experiment for the characterization of silicophosphate crystalline phases and amorphous gels, Inorg. Chem., 2007, 46, 1379–1387.

[24] Elena, B., Lesage, A., Steuernagel, S., Bockmann, A., & Emsley, L. Proton to carbon-13 INEPT in solid-state NMR spectroscopy, J. Am. Chem. Soc., 2005, 127, 17296–17302.

[25] Bax, A., & Summers, M. F. Proton and carbon-13 assignments from sensitivity-enhanced detection of heteronuclear multiple-bond connectivity by 2D multiple quantum NMR, J. Am. Chem. Soc., 1986, 108, 2093–2094.

[26] Lesage, A., & Emsley, L. Through-bond heteronuclear single-quantum correlation spectroscopy in solid-state NMR, and comparison to other through-bond and through-space experiments, J. Magn. Reson., 2001, 148, 449–454.

[27] Wen, J. L., Sun, S. L., Xue, B. L., & Sun, R. C. Recent advances in characterization of lignin polymer by solution-state nuclear magnetic resonance (NMR) methodology, Materials, 2013, 6, 359–391.

[28] Dorkoosha, F.A., Brusseeb, J., Verhoeff, , Borcharda, G., Tehrania, M. R., & Junginger, H. E. Preparation and NMR characterization of superporous hydrogels (SPH) and SPH composites, Polymer, 2000, 41, 8213–8220.

[29] Bertmer, M., Buda, A., Höfges, I. B., Kelch, S., & Lendlein, A. Biodegradable shape-memory polymer networks: Characterization with solid-state NMR, Macromolecules, 2005, 38, 3793–3799.

[30] Gussoni, M., Greco, F., Mapelli, M., Vezzoli, A., Ranucci, E., Ferruti, P., & Zetta, L. Elastomeric polymers. 2. NMR and NMR imaging characterization of cross-linked PDMS, Macromolecules, 2002, 35, 1722–1729.

[31] Río, J. C. D., Rencoret, J., Prinsen, P., Martínez, A. T., Ralph, J., & Gutiérrez, A. Structural characterization of wheat straw lignin as revealed by analytical pyrolysis, 2D-NMR, and reductive cleavage methods, J. Agric. Food Chem., 2012, 60, 5922–5935.

[32] Agarwal, G. K., & Titman, J. J. Characterization of vinyl polymer/silica colloidal nanocomposites using solid-state NMR spectroscopy: Probing the interaction between the inorganic and organic phases on the molecular level, J. Phys. Chem. B, 2003, 107, 12497–12502.

[33] Xu, J-Z., Ma, R., Jiao, Y.H., Xie, J.X., Wang, R., & Su, L. Synthesis, characterization, and thermal degradation property of aniline polyphosphazene, J. Macromol. Sci., Part B: Phys., 2011, 50(5), 897–906.

[34] Shuangkun, Z., Ali, S., Ma, H., Zhang, L., Wu, Z., & Wu, D. Preparation of poly(bis(phenoxy) phosphazene) and 31P NMR analysis of its structural defects under various synthesis conditions, J. Phys. Chem. B, 2016, 120(43), 11307–11316.

[35] Boccaccio, T., Bottino, A., Capannelli, G., & Piaggio, P. Characterization of PVDF membranes by vibrational spectroscopy, J. Memb. Sci., 2002, 210, 315–329.

[36] Conley, R. T. Thermal stability of polymers, Marcell Dekker, New York, 1973.

[37] Wang, Q., & Shi, W. Kinetics study of the thermal decomposition of epoxy resins containing flame retardant components, Polym. Deg. Stab., 2006, 91, 1747–1754.

[38] Laza, J. M., Vilas, J. L., Garay, M. T., Rodríguez, M., & León, L. M. Dynamic mechanical properties of epoxy-phenolic mixtures, J. Polym. Sci., Part B: Polym. Phys., 2005, 43, 1548–1557.

[39] Barontini, F., Marsanich, K., Petrarcaca, L., & Cozzani, V. Thermal degradati, n and decomposition products of electronic boards containing BFRs, Ind. Eng. Chem. Res., 2005, 44, 4186–4193.

[40] Heisenberg, E. Cellulose Chemie, 1931, 12, 159. C.A. 25, 59, 823, 1931.

[41] Duswalt, A. A. The practice of obtaining kinetic data by differential scanning calorimetry, Therm. Acta, 1974, 8, 57–68.

[42] Levi, D. W., Reich, L., & Lee, H. T. Degradation of polymers by thermal gravimetric techniques, Polym. Engg. Sci., 1965, 5, 135–141.

[43] Friedman, H. L. U. S. Dept. Com., Office. Tech. 24 pp, 1959, Chem. Abstr., 1961, 55(26), 511.

[44] Reich, L., & Levi, D. W. Macromol. Rev, Eds, Peterlin Goodman Wiley Interscience, New York, 1968, 173.

[45] Xie, W., Gao, , Liu, K., Pan, W. P., Vaia, R., Hunter, D., & Singh, A. Thermal characterization of organically modified montmorillonite, Therm. Acta, 2001, 367–368, 339–350.

[46] Carrascoa, F. P. P., Pérez C, J. G., Santanac, O. O., & Maspoch, M. L. Processing of poly(lactic acid): characterization of chemical structure, thermal stability and mechanical properties, Polym. Degrad. Stab., 2010, 95, 116–125.

[47] Cerqueira, M. A., Souzaa, B.W.S., Simões, J., Teixeira, J. A., Domingues, M. R. M., Coimbra, M. A., & Vicente, A. A. Structural and thermal characterization of galactomannans from non-conventional sources, Carb. Polym., 2011, 83, 179–185.

[48] Verploegen, E ., Mondal, R., Bettinger, C. J., Sok, S., Tone Y, M. F., & Bao, Z. Effects of thermal annealing upon the morphology of polymer-fullerene blends, Adv. Funct. Mater., 2010, 20, 3519–3529.

[49] Kübel, C., González-Ronda, L., Drummy, L. F., & Martin, D. C. Defect-mediated curvature and twisting in polymer crystals, J. Phys. Org. Chem., 2000, 13, 816–829.

[50] Reiter, G. Some unique features of polymer crystallization, Chem. Soc. Rev., 2014, 43, 2055–2065.

[51] Strobl, G. Crystallization and melting of bulk polymers: new observations, conclusions and a thermodynamic scheme, Prog. Polym Sci., 2006, 31, 398–442.

[52] Agbolaghi, S., Abbaspoor, S., & Abbasi, F. A comprehensive review on polymer single crystal from fundamentalal concepts to applications, Prog. Polym. Sci., 2018, 81, 22–79.

[53] Zachmann, H. G. Theory of nucleation and crystal growth of polymers in concentrated solutions, Pure Appl. Chem., 1974, 38, 79–96.

[54] Ma, M., He, Z., Yang, J., Chen, F., Wang, K., Zhang, Q. et al.. Effect of film thickness on morphological evolution in dewetting and crystallization of polystyrene/poly((-caprolactone) blend films, Langmuir, 2011, 27, 13072–13081.

[55] Li, H., & Yan, S. Surface-induced polymer crystallization and the resultant structures and morphologies, Macromolecules, 2011, 44, 417–428.

[56] Cheng, S. Z. D. Phase transitions in polymers: the role of metastable states, Amsterdam:, Elsevier Ltd, 2008, 81–142.

[57] Shcherbina, M. A., & Ungar, G. Analysis of crystal habits boundeby asymmetrically curved faces: polyethylene oligomers and polvinylidene fluoride), Polymer, 2007, 48, 2087–2097.

[58] Shcherbina, M. A., & Ungar, G. Asymmetric curvature of growth faces copolymer crystals, Macromolecules, 2007, 40, 402–405.

[59] Gestí, S., Almontassir, A., Casas, M. T., & Puiggalí, J. Crystalline structure of poly (hexamethylene adipate): study on the morphology and the enzymatic degradation of single crystals, Biomacromolecules, 2006, 7, 799–808.

[60] Gestí, S., Almontassir, A., Casas, M. T., & Puiggalí, J. Molecular packing and crystalline morphologies of biodegradable poly(alkylene dicarboxylate)s derived from 1,6-hexanediol, Polymer, 2004, 45, 8845–8861.

[61] Gestí, S., Casas, M. T., & Puiggalí, J. Crystalline structure of poly(hexamethylene succinate) and single crystal degradation studies, Polymer, 2007, 48, 5088–5097.

[62] Weber, C. H. M., Chiche, A., Krausch, G., Rosenfeldt, S., Ballauff, M., & Harnau, L. Single lamella nanoparticles of polyethylene, Nano. Lett., 2007, 7, 2024–2029.

[63] Su, M., Huang, H., Ma, X., Wang, Q., & Su, Z. Poly(vinyl pyridinene)-block-poly(ε-caprolactone) single crystals in micellar solution, Macromol. Rapid Commun., 2013, 34, 1067–1071.

[64] Ungar, G., Putra, E. G. R., De Silva, D. S. M., Shcherbina, M. A., & Waddon, A. J. The effect of self-poisoning on crystal morphology and growth rates, Adv. Polym. Sci., 2005, 180, 45–87.

[65] Sasaki, S., & Asakura, T. Helix distortion and crystal structure of the α-form of poly(l-lactide), Macromolecules, 2003, 36, 8385–8390.

[66] Aleman, C., Lotz, B., & Puiggali, J. Crystal structure of the -form of poly(l-lactide), Macromolecules, 2001, 34, 4795–4801.

[67] Ungar, G., & Putra, E.G.R. Asymmetric curvature of {110} crystal growth face in polyethylene oligomers, Macromolecules, 2001, 34, 5180–5185.

[68]  Lauritzen, J. I., & Hoffman, J. D. Theory of formation of polymer crystals with folded chains in dilute solution, J. Res. Natl. Bur. Stand., 1960, 64A, 73–102.

[69]  Lauritzen, J. I., & Hoffman, J. D. Extension of theory of the growth of chain-folded polymer crystals to large undercoolings, J. Appl. Phys., 1973, 44, 4340–4352.

[70]  Muthukumar, M., & Welch, P. Modeling polymer crystallization from solutions, Polymer, 2000, 41, 8833–8837.

[71]  Welch, P., & Muthukumar, M. Molecular mechanisms of polymer crystallization from solution, Phys. Rev. Lett., 2001, 87(218302), 1–4.

[72]  Yao, Y. F., Graf, R., Spiess, H. W., Lippits, D. R., & Rastogi, S. Morphological differences in semi-crystalline polymers: implications for local dynamics and chain diffusion, Phys Rev. E, 2007, 76(060801), 1–4.

[73]  Chen, R., Li, L., & Zhao, J. Single chain diffusion of poly(ethylene oxide) in its monolayers before and after crystallization, Langmuir, 2010, 26, 5951–5956.

[74]  Liu, Y. X., Li, J. F., Zhu, D. S., Chen, E. Q., & Zhang, H. D. Direct observation and modeling of transient nucleation in isothermal thickening of polymer lamellar crystal monolayers, Macromolecules, 2009, 42, 2886–2890.

[75]  Luo, C., & Sommer, J. U. Growth pathway and precursor states in single lamellar crystallization: MD simulations, Macromolecules, 2011, 44, 1523–1529.

# Index

https://doi.org/10.1515/9781501514609-010